Joaquín M. Campos Rosa
Drug Design and Action

Also of interest

Pharmaceutical Chemistry.
Volume 2: Drugs and Their Biological Targets
Joaquín M. Campos Rosa, planned 2024
ISBN 978-3-11-131655-0, e-ISBN 978-3-11-131688-8

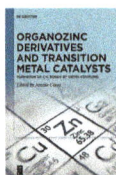

Organozinc Derivatives and Transition Metal Catalysts.
Formation of C-C Bonds by Cross-coupling
Janine Cossy (Ed.), 2023
ISBN 978-3-11-072877-4, e-ISBN 978-3-11-072885-9

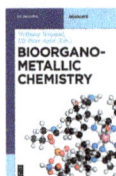

Bioorganometallic Chemistry
Wolfgang Weigand, Ulf-Peter Apfel (Eds.), 2020
ISBN 978-3-11-049650-5, e-ISBN 978-3-11-049657-4

Chemical Drug Design
Girish Kumar Gupta, Vinod Kumar (Eds.), 2016
ISBN 978-3-11-037449-0, e-ISBN 978-3-11-036882-6

Joaquín M. Campos Rosa

Pharmaceutical Chemistry

Volume 1: Drug Design and Action

2nd Edition

DE GRUYTER

Author
Prof. Joaquín M. Campos Rosa
Department of Pharmaceutical and Organic Chemistry
University of Granada
Campus de Cartuja
18071 Granada
Spain
jmcampos@ugr.es

ISBN 978-3-11-131654-3
e-ISBN (PDF) 978-3-11-131690-1
e-ISBN (EPUB) 978-3-11-131721-2

Library of Congress Control Number: 2023946629

Bibliographic information published by the Deutsche Nationalbibliothek
The Deutsche Nationalbibliothek lists this publication in the Deutsche Nationalbibliografie;
detailed bibliographic data are available on the internet at http://dnb.dnb.de.

© 2024 Walter de Gruyter GmbH, Berlin/Boston
Cover image: alice-photo/iStock/Getty Images Plus
Typesetting: Integra Software Services Pvt. Ltd.
Printing and binding: CPI books GmbH, Leck

www.degruyter.com

"The expert in anything was once a beginner"
Anonymous

"What we learn with pleasure we never forget"
Alfred Mercier (1816–1894)

Preface to the first edition

Global objectives of pharmaceutical chemistry

1. To understand the interrelation between structure, physicochemical properties, pharmacological activity, and therapeutic utility
2. To know the methods and strategies used in the generation of drugs
3. To know the interactions between drugs and their biological targets
4. To know and propose the structural modifications that affect the properties of the drugs
5. To know the general methods and the synthetic strategies for the preparation of drugs
6. To know the analytical and spectroscopic methods applicable to the structural identification and elucidation of drugs, and related compounds
7. To be able to name and formulate a drug in accordance with the systematic IUPAC nomenclature
8. To know and become able to predict the transformation of drugs in the body
9. To know and be able to estimate the risks associated with the use of reagents, solvents and the development of processes in the chemical laboratory
10. To know how to acquire and use information related to drugs

The objective of pharmaceutical chemistry (hereinafter PC) is the chemical study of drugs and the active ingredients of drugs in order to determine the relationship between chemical structure, physicochemical properties, reactivity, and biological response, with the ultimate aim of providing the knowledge necessary for the creation of new drugs. Since most of the drugs are organic in nature, PC is mainly based on the knowledge of organic chemistry, although it requires a strong familiarity with and a solid base in biochemistry. It is also nourished by other sciences such as (a) pharmacognosy, which studies natural products as a source of new active principles; (b) pharmacology, which allows the establishment of experimental models for the evaluation of new active principles; and (c) molecular pharmacology. The latter tries to explain the biological effects at the molecular level, interpreting the phenomena related to the association between a drug and the biomolecules that trigger its action, all from the point of view of structural and physicochemical properties.

Although drug design in its origins focused primarily on the simple chemical modifications of naturally occurring molecules, current design trends are based on the study of drug interactions with their targets at the molecular level. The development of molecular biology and genetic engineering over the last decade has allowed the detailed study of many targets in the action of drugs, such as enzymes, membrane receptors, and nucleic acids. Therefore, part of the design of new drugs is currently based on drug–receptor interactions.

https://doi.org/10.1515/9783111316901-202

The complete work consists of the following two volumes:

Pharmaceutical Chemistry
Volume 1: Drug Design and Action

Pharmaceutical Chemistry
Volume 2: Drugs and Their Biological Targets

After the first part, in which general principles are explained, in the second volume, the knowledge acquired for the establishment of the therapeutic arsenal according to the different molecular objectives is applied. There are excellent treatises of PC or medicinal chemistry, but require the reader to have a solid foundation. Our book is aimed at pharmacy and chemistry students who intend to enter the exciting world of drug development. Therefore, we have not tried to cover this study in an exhaustive way, but rather to establish the bases that in a first stage allow the fostering of interest in this scientific field. This is our humble goal, and if we achieve it, we will feel satisfied that we have fully achieved our objective. We have tried to achieve a balance between chemical and biological aspects, highlighting the strong multidisciplinary character of this science. As far as possible, the number of drugs of both volumes has been reduced to a minimum, sufficient to understand the philosophy of PC and without overburdening the beginner with more and more examples, and therefore we hope to avoid a "can't see the wood for the trees".

Drug synthesis is approached from a double point of view in Chapter 8 of this first volume:

(a) The retrosynthetic analysis that allows the bond breaking of the molecule to arrive at a structurally simpler starting material and that will allow us to carry out the direct and real process of synthesis in a rational way.

(b) The ability to propose the corresponding synthesis, based on the information obtained in the retrosynthetic analysis, mechanism of reactions, and reactivity of organic compounds. The basis of biology is chemistry, which permits the creation of drugs (chemists acting as molecular architects), as well as the analysis of the interactions between the drug and its biological objective, which will permit the design of more active structures after their optimization. This first volume provides exercises and their solutions, allowing the reader to assess to what degree he or she has understood

The work shown here is the result of extensive teaching experience at the Faculty of Pharmacy of Granada University. We want to thank the role of the students in the noble task of teaching. They give meaning to our work and stimulate us to keep both knowledge and methodology up to date. We want to thank our parents, partners, and children for the support we have received. They have demonstrated that constancy is fundamental to achieving our objectives, while we have neglected our families during the time devoted to this work.

It is the explicit desire of the authors to receive any suggestions, additions and corrections, which will surely make possible an expansion of the contents of the work.

Fundamental bibliography

Victoria F. Roche, S. William Zito, Thomas L. Lemke, David A. Williams. Foye's Principles of Medicinal Chemistry. Wolters Kluwer Heath, 8th Edition, 2019.

Thomas Nogrady, Donald F. Weaver. Medicinal Chemistry. A Biochemical Approach. Editorial Oxford University Press, 3rd Edition, 2005.

Patrick, Graham L. An Introduction to Medicinal Chemistry. Editorial Oxford University Press, 5th Edition, 2013.

Enrique Raviña. The Evolution of Drug Discovery: From Traditional Medicines to Modern Drugs. Wiley-VCH, Verlag GmbH & Co., KGaA, 1st Edition, 2011.

Richard B. Silverman, Mark W. Holladay. The Organic Chemistry of Drug Design and Drug Action. Elsevier Academic Press, 3rd Edition, 2015.

Camille Georges Wermuth, David Aldous, Pierre Raboisson, Didier Rognan. The Practice of Medicinal Chemistry. Academic Press, 4th Edition, 2015.

Corwin Hansch, Albert Leo. Exploring QSAR: Fundamentals and Applications in Chemistry and Biology. American Chemical Society, Washington, DC, 1995.

Hugo Kubinyi. QSAR: Hansch Analysis and Related Approaches, VCH Verlagesellschaft mbH, 1993.

Complementary bibliography

Daniel Lednicer. Organic Chemistry of Drug Synthesis. Vols. 1–6, Wiley-Interscience, 1977–1998.

Stuart Warren, Paul Wyatt. Organic Synthesis: The Disconnection Approach. Wiley, 2nd Edition, 2008.

Preface to the second edition

The objective of pharmaceutical chemistry (PC) is the chemical study of drugs, the active ingredients of medicines, in order to determine the relationship between chemical structure, physicochemical properties, reactivity, and biological response, with the ultimate aim of providing the necessary knowledge for the creation of new drugs.

Since most drugs are organic in nature, therapeutic chemistry or PC is mainly based on the knowledge of organic chemistry, although it requires a strong biological connection with a solid foundation in biochemistry. On the other hand, it also draws on other subjects such as,

a) Pharmacognosy, which studies natural products as a source of new active ingredients.
b) Pharmacology, which allows the establishment of experimental models for the evaluation of new active ingredients.
c) Molecular pharmacology, which tries to explain biological effects at the molecular level, interpreting the phenomena related to the association between a drug and the biomolecules that trigger its action, all from the point of view of structural and physicochemical properties.

Although drug design, the ultimate objective of PC, was originally focused on simple chemical modifications of molecules of natural origin, current design trends are based on the study of drug interactions with their target structures at the molecular level. The development experienced in the last decade by molecular biology and genetic engineering has allowed the detailed study of many target molecules in drug action, such as enzymes, membrane receptors and nucleic acids. Therefore, part of the design of new drugs is currently based on drug–receptor interactions.

The synthesis of the designed compounds is another aspect to be considered in the study of PC.

Organic chemistry is approached from a mechanistic viewpoint. The ability to write feasible reaction mechanisms in organic chemistry depends on the extent of the individual's preparation. These two volumes assume the knowledge obtained in a two-semester undergraduate course of organic chemistry. I am going to review how to draw Lewis structures and how to calculate formal charges. In order to write or understand reaction mechanisms, it is essential to be able to construct Lewis structures for any organic compound. The electron flow pathway has become the building blocks of the mechanistic processes. I am going to use the concept of electron flow with the rigorous use of curved arrows as an electron bookkeeping device.

I have attempted to keep the use of IUPAC names of organic compounds to a bare minimum to enhance the readability of the text. In the discussion of reactions, I drew out the structures rather than the names of the compounds.

I will try to condense the essentials of PC in a student-friendly manageable way. I will do this by concentrating purely on the basics of the subject without going into

https://doi.org/10.1515/9783111316901-203

exhaustive detail or repetitive examples. Furthermore, keynotes at the end of a topic summarize the essential facts covered and help to focus the mind on the essentials. On the other hand, the most important aspects of some topics will be indicated at the end of the corresponding chapter and will constitute the skeleton of the topic. They are like the branches of a bare Christmas tree to which the leaves and ornaments will have to be added.

After so many years dedicated to research and teaching, I have learned to question everything!

Joaquín M. Campos Rosa
September 2023

How to study pharmaceutical chemistry effectively

- **Ask for help**. Your professors have office hours that are specifically available for students to ask questions and get help with the material. Take advantage of this time, go and ask for assistance. They are there to help you learn and this is an opportunity to ask questions you might have been afraid to ask in class. **Getting extra help from a one-to-one tutorship can also be beneficial if you are really struggling.**
- **Study every day**. Studying for an hour or two each day is a much better use of your time than trying to study for 5 h all at once. **With smaller, more frequent study sessions, you can focus on individual concepts and not burn yourself out trying to cram all of the information into your brain at once**.
- **Attend class**. Class attendance is essential because it facilitates the first under-standing of what is explained and makes learning more pleasant. Paying attention in class is also necessary. Listen to what the professor is saying and then summarize the information in your notes.
- **Understanding the how and why of a biological, organic reaction or concept will lead to a higher score**. Many scores start high but can fall gradually through-out the semester. These are students who failed to put in the effort after the mid-term exam (overconfident) or who relied on memorizing the material rather than understanding and building on a concept.
- **Do not procrastinate (to put off or defer an action until a later time; delay)**. Do something refreshing, because all studying without a break can make it diffi-cult to concentrate.
- **PC and organic chemistry are amazing subjects because of their hierarchical nature. Learn a few simple concepts and you can explain and solve many problems throughout the year.**
- **Getting behind early is hard to overcome but it is not a crime**. If you get a bad grade in your midterm exam, you must learn from your mistake and start again. Many students have failed the first exam of the course and gone on to get an A as a final grade. **Be strong and constant enough!**

https://doi.org/10.1515/9783111316901-204

How to achieve success in exams: some specific hints for pharmaceutical chemistry

Chemistry forms the building blocks of life. It is a subject traditionally viewed as difficult because of the complex nature of chemical reactions and the unfamiliarity of the new language.

– **Tip 1:** The secret to success is using formulae! A simple mistake like failing to draw a formula, can affect your marks greatly!
– **Tip 2: When it comes to answering theoretical questions, the explanation must be based on chemical structures, mechanisms, or biological schemes.** However, all these previous forms must be verbalized and justified. Both extremes are not valid: (a) representation of schemes, structures, and so on without providing a coherent explanation or (b) a lot of text but without the commitment of structural representation of the corresponding cartoon. The professor has to extract from the exam the global content of your training and not imagine what you mean.
– **Tip 3: Be organized, structured, and clean** (without crossing out) when solving an exam. The more the teacher's work is facilitated, the better the test grade will be.

Ultimately, you should walk away from chemistry and understand more about the world around you.

Chemistry is interesting and fun, but it takes time, effort, and dedication.

Do not be afraid to ask for help, many people have been where you are right now and pulled through!

Alongside the improvement in your academic results, the most outstanding and important outcome would be the enhancement of your self-motivation, and **even more importantly, development of your self-confidence.**

> *"For my part, I believe that medicines are one of the blessings of our age, perhaps the greatest of them all"*
>
> Sir Ernst Boris Chain (1906–1979)

> Co-recipient of the Nobel Prize in Physiology and Medicine for his work on penicillin.

> *"Success consists of going from failure to failure without loss of enthusiasm"*
>
> Sir Winston Churchill (1874–1965)

https://doi.org/10.1515/9783111316901-205

Contents

About the author

Joaquín M. Campos received his B.Sc. and Ph.D. from the University of Granada (UGR). He is full professor in medicinal and organic chemistry. He worked with Professor Ganellin, coinventor of cimetidine (Tagamet®), at University College London, UK. He has published 180 papers in international journals with a high impact index, mainly in the fields of organic chemistry, medical chemistry, and cancer; 14 books; 14 book chapters; 13 national and 17 international patents; and supervised 15 doctoral theses. He has been involved for the most part as a responsible researcher in 18 research projects and 3 research contracts with private companies. He advised the University of São Paulo (Brazil) on novel anticancer and anti-Alzheimer's drugs, during the years 2015–17 as special visiting professor. He was awarded the 2010 Teaching Excellence Award from UGR. On March 17, 2016, he received the award from the Social Council of UGR. He is regional editor in Europe for the journal *Current Medicinal Chemistry* since August 13, 2016. *Current Medicinal Chemistry* is a leading journal in the field of medicinal chemistry.

As a consequence of his interest in raising the level of internationalization of the UGR and more specifically that of the Faculty of Pharmacy of the UGR, he has been teaching pharmaceutical chemistry in the Pharmacy Degree in English for many years, with excellent acceptance among students.

https://doi.org/10.1515/9783111316901-207

1 Basic concepts in pharmaceutical chemistry

1.1 Goals

- To know the tasks of pharmaceutical chemistry
- To know and understand the concepts of drugs and medicines
- To understand the multidisciplinary nature of this discipline
- To introduce the concepts of patent and pharmaceutical industry
- To know the global process through which new drugs are discovered and the methodologies used over time
- To know the main stages of development of a new drug and its cost

1.2 Basic concepts and purposes of pharmaceutical chemistry

According to the IUPAC specialized commission, pharmaceutical chemistry (PC) is related to the discovery, development, identification, and interpretation of the mechanism of action of biologically active compounds. Drugs will be emphasized, but the interest of pharmaceutical chemists will not be restricted only to drugs, but also include biologically active compounds, in general. PC will also address the study, identification, and synthesis of the metabolic products of drugs and related compounds. In short, PC is responsible for the design, synthesis, and analysis of drugs. Scheme 1.1 shows the general method of work in PC.

After the application of Scheme 1.1, the ideal new drug has to fulfill the following conditions:

1. New patentable chemical entity capable of being registered
2. Its synthetic process should not exceed four steps and should not include any heavy metal catalysts or environmentally problematic wastes; purity >99%
3. Stable up to 70 °C, even in a humid environment, and stable against light
4. It has to possess solid-state properties (crystalline, nonexistence of polymorphic and non-hygroscopic forms)
5. Solubility in water
6. 90% oral bioavailability
7. High activity with a pharmacokinetic profile that allows it to be given once daily with a dose

https://doi.org/10.1515/9783111316901-001

PHARMACEUTICAL CHEMISTRY

It is responsible for the design, synthesis and analysis of drugs

IDENTIFY THE CHEMICAL STRUCTURE FOR SYNTHESIS

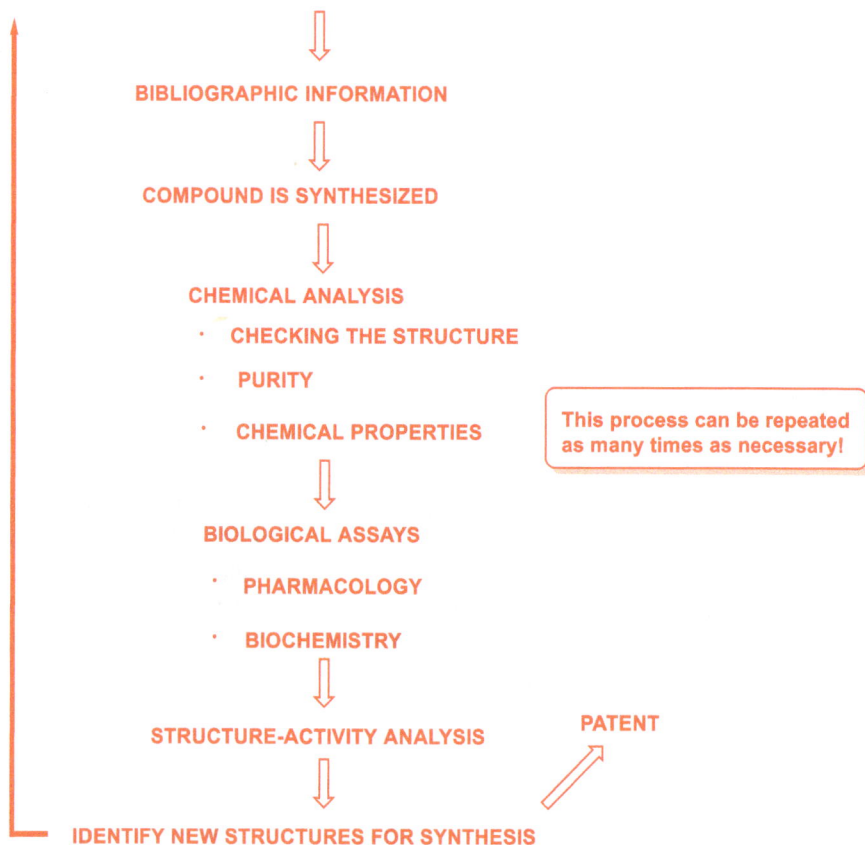

⇓

BIBLIOGRAPHIC INFORMATION

⇓

COMPOUND IS SYNTHESIZED

⇓

CHEMICAL ANALYSIS

· **CHECKING THE STRUCTURE**

· **PURITY**

· **CHEMICAL PROPERTIES**

This process can be repeated as many times as necessary!

⇓

BIOLOGICAL ASSAYS

· **PHARMACOLOGY**

· **BIOCHEMISTRY**

⇓

STRUCTURE-ACTIVITY ANALYSIS   **PATENT**

⇓

IDENTIFY NEW STRUCTURES FOR SYNTHESIS

Scheme 1.1: The general method of work in PC.

PC has three fundamental objectives:
1. Structural modification of compounds that have a well-known physiological action, i.e. obtaining new drugs from other ones already known: molecular manipulation.
2. It provides the necessary knowledge for the development of new drugs: drug design.
3. In addition, the use of very sensitive analytical techniques is essential for the quality control that determines the efficacy and safety of the medicines: drug analysis. On the other hand, metabolite analysis is a key aspect in the pharmacokinetic studies that determine the bioavailability and duration of the therapeutic response. Purity tests set the limit for acceptable impurities. If the tests have to determine the amount of an active substance in a medicinal product, it must be separated from other accompanying substances by applying separation techniques.

There are definitions that should be known:

Drug: any raw material of animal or vegetable origin that contains one or several active principles which, introduced into the body by any route of administration, produces an alteration of the natural functioning of the central nervous system of the individual and is also susceptible of creating dependence, whether psychological, physical, or both. In English, a drug is also a biologically active substance (capable of interacting with the biological environment), chemically pure, and with therapeutic action, i.e. capable of curing, mitigating, or preventing disease in humans or animals (e.g. acetylsalicylic acid).

Medicine: the drug in the proper dosage form, used in medicine (e.g. Aspirin®, Fig. 1.1). It is synonymous with drug. Hence, the terms "medicinal chemistry", "therapeutic chemistry", or "pharmaceutical chemistry" are used interchangeably.

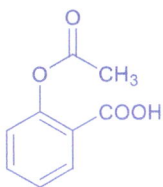

Fig. 1.1: Acetylsalicylic acid (Aspirin®).

The mere disposition of active drugs is not sufficient for their clinical application: it is necessary to formulate medicines that can be introduced into the organism by the appropriate route for each patient. For example, acetylsalicylic acid, widely used as an analgesic, antipyretic, blood thinner, and an antirheumatic drug, could not be used without a previous physicochemical transformation which would allow its absorption and, if necessary, neutralize its ulcerogenic properties, contraindicated for certain patients. It has been demonstrated that the gastrointestinal dissolution of acetylsalicylic acid is the slowest step, and is the determinant of its rate of absorption; therefore, several forms of oral administration are required:

1. Micronization of its crystalline form to an ultrafine powder of high dissolution rate, as in tablets known under the trade name of Aspirin®
2. Its galenic formulation with antacids or buffer substances, to form locally soluble acetylsalicylates (effervescent Aspirin®)
3. The use of enteric-coated tablets that avoid direct contact of the drug with the gastric mucosa, even if it is detrimental to absorption

We will study drugs (pure substances) and not medicines, in this collection.

1.3 Historical development of pharmaceutical chemistry

We can distinguish three great periods:

1.3.1 Pre-scientific period (~ 3000 years BC until the nineteenth century)

In ancient times, scientists were philosophers and erudite people so that the different sciences were purely empirical. Some of the natural products, either as such or in the form of derivatives, were often used for various purposes, such as arrow poisons, complements for religious rituals, or even cosmetics:

Belladonna: Today   → Antimuscarinic drug
            Previously   → Cosmetic or poison

Curare: Today   → Muscle relaxant
            Previously   → Poison

1.3.2 Scientific period (nineteenth century to 1960)

From the nineteenth century and after the French Revolution (at the end of the eighteenth century), the sciences went from being empirical to being sciences based on experimentation. Organic chemistry was developed and, as a consequence, synthetic products began to be used. The first occasion in which a synthetic organic product was used to interfere with vital processes was probably during the first half of the nineteenth century when ether and chloroform were introduced as anesthetics. Because of this, initial efforts to search for new synthetic drugs focused primarily on anesthetics and hypnotics, and subsequently on analgesics. Physiology was also developed, and most diseases were listed and classified. A key figure in this period is the German immunologist Paul Ehrlich (1854–1915), who thought that drugs were able to distinguish human cells from those of parasites. This assumption was reinforced by his previous experiences on the selective staining of different tissues of mammals with dyes, as well as his studies on the selectivity of antibodies to the corresponding antigens (substances that cause the formation of antibodies when introduced into the body). He is the father of chemotherapy, which he defined as the use of drugs that harm the invading organism without causing harm to the host. This greatest contribution to the advancement of PC is probably the original ideas that he proposed about the mode of action of the drugs. Thus, he postulated the existence of receptors in mammalian cells (a receptor is a macromolecule to which various ligands or compounds selectively bind that cause a specific biological effect). He distinguished two parts in antigens and chemotherapeutic agents:

1. Haptophore groups: responsible for the union
2. Toxoplastic groups: responsible for toxicity

1.3.3 Current period (from the 1960s to the present)

With the development of biochemistry and molecular pharmacology, the receptors and the mechanism of action of the drugs began to be known, and so it is logical to think of a structure–activity relationship (SAR). If it is accepted that the activity of a drug is due to its chemical interaction with a hypothetical cellular receptor, it is logical to consider that its physicochemical properties, and therefore, its structure, must be directly related to its activity. In 1964, the two most solid and general procedures of the quantitative SAR appear:
– Hansch–Fujita method
– Free–Wilson method

Nowadays, with these and other theoretical studies, drugs can be rationally designed.

1.4 Multidisciplinary nature of this discipline

For the development of PC, a solid chemical knowledge, especially of organic chemistry, is not enough, but requires a strong biological base, concretized in a rational foundation of biochemistry.

Molecules that show biological activity are called hits. The next step is to find compounds that have attractive pharmaceutical properties, including low toxicity, aqueous solubility suitable for being orally administered, among other pharmacokinetic properties. Such compounds are called "leaders or leads". Leaders are hits but more refined. Typically, hits are found by screening a vast number of molecules, while "leader compounds" are developed from hits through chemical modifications.

Hit to lead, also known as lead generation, is a stage in early drug discovery where small molecule hits from a high-throughput screening are evaluated and undergo limited optimization to identify promising lead compounds. These lead compounds undergo more extensive optimization in a subsequent step of drug discovery called lead optimization. On the other hand, a lead is a compound or a series of compounds with proven activity and selectivity that meets the criteria for drug development such as originality, patentability, and accessibility (by extraction or synthesis). PC relies on (a) pharmacognosy, which studies drugs as a source of active principles – such active principles themselves constitute authentic drugs and serve as models for obtaining new drugs by molecular manipulation; and (b) pharmacology and pharmacodynamics, which study the action of drugs and their mechanism. Having checked the pharmacological action of a drug, its pharmaceutical form allowing its administration to the patient should be given. The adequate drug formulation requires biopharmaceutical studies (i.e. those factors that influence the bioavailability of drugs) and pharmacokinetics, which refer to the kinetics of absorption, distribution, metabolism, and excretion of the drugs and their therapeutic response or toxicity in animals and man. Biopharmacy and pharmacokinetics will therefore complete the cycle of the drug when introduced into the galenic formulation (Fig. 1.2).

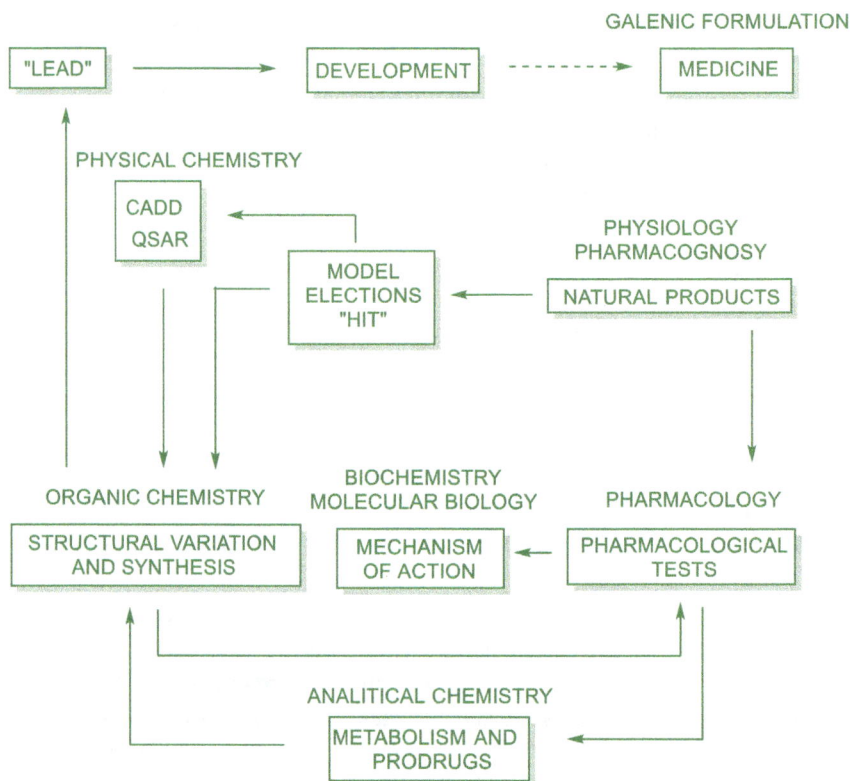

Fig. 1.2: Relationships of the different disciplines that affect PC.

1.5 Patents

A patent is a title that grants the right to manufacture and market the object of the patent during the period of validity (which is usually 20 years). That which is patented must be novel, implying that it has not been disclosed, and must have an inventive step, that is, it should not be obvious to a person skilled in the art. One of the hot topics in the patent field is the "chiral switch". During the period from 1983 to 1987, 30% of the approved drugs were pure enantiomers, 29% were racemic, and 41% were achiral. Nowadays, most of the drugs that reach the market are achiral or pure enantiomers. The racemic problem is that each of the enantiomers normally has a different level of activity. In addition, the enantiomers may differ in the way in which they are metabolized and in their side effects. Therefore, it is better to market a pure enantiomer rather than a racemic one. The issue of chiral change is mostly related to racemic drugs that have been on the market for several years and for which the 20-year patent is on the point of expiring. Through the *chiral switch* toward a pure enantiomer, the pharmaceutical companies can argue that it is a new invention and that therefore a new patent can be formalized. However, they have to prove that the pure enantiomer represents an improvement over the original racemic, and such a fact was not predictable when the racemic was originally patented. Examples of drugs for which chiral change has been performed are salbutamol and omeprazole (Fig. 1.3).

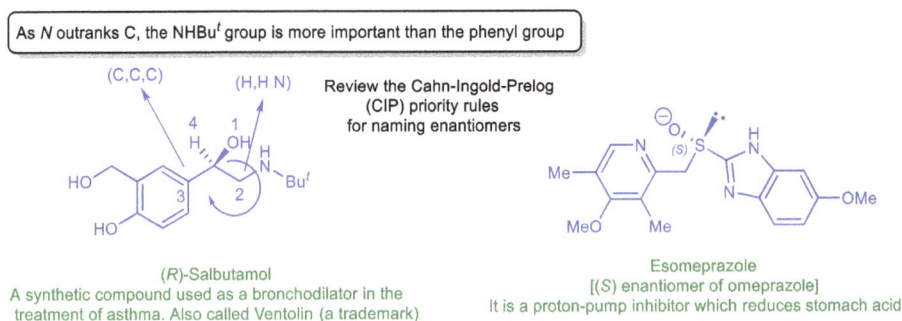

Fig. 1.3: The chiral switch has been carried out in some drugs.

1.5.1 Cahn–Ingold–Prelog priority rules

- Compare the atomic number (Z) of the atoms directly attached to the stereocenter: the group having the atom of higher atomic number receives higher priority.
- If there is a tie, we must consider the atoms at distance 2 from the stereocenter: the group containing the atom of higher atomic number receives higher priority.
- If there is still a tie, we must consider the atoms at distance 3 from the stereocenter), and classified in decreasing order of atomic number, and so on.
- If an atom A is double bonded to an atom B, A is treated as being singly bonded to two atoms.
- After the substituents of a stereocenter have been assigned their priorities, the tetrahedron is oriented in space so that the group with the lowest priority is pointed away from the observer. You have to look at the tetrahedron through the lowest priority group. If the substituents are numbered from 1 (highest priority) to 4 (lowest priority), then the sense of rotation passing through 1, 2, and 3 distinguishes the

stereoisomers. A center with a clockwise sense of rotation is an *R* (*rectus*) center and a center with a counterclockwise sense of rotation is an *S* (*sinister*) center.

1.5.2 Importance of homochiral drug development

–   The introduction of racemic drugs is becoming less attractive due to the new policies followed by various regulatory agencies. Therefore, the preparation of pure enantiomeric drugs (homochiral) is a critical issue of the pharmaceutical industry. There are different reasons for producing optically pure materials:
–   **(a)** Biological activity can be associated with a single enantiomer. In the most general situation, the enantiomers can exhibit different types of activities, of which both can be beneficial, or one can be and the other undesirable. The production of a single enantiomer allows, therefore, the separation of the effects.
–   **(b)** The unwanted isomer is an "isomeric ballast" exposed to the surrounding environment. The production of the active enantiomer is a matter of law in certain countries, being considered the unwanted enantiomer as an impurity.

1.5.2.1 The thalidomide tragedy
Thalidomide (Fig. 1.4), a derivative of glutamic acid, was introduced as a racemic in therapeutics in the 1960s as a sedative-hypnotic agent and was used to alleviate the nausea of pregnant women.

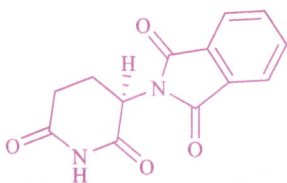

Fig. 1.4: (*R*)-(+)-Thalidomide of α-(*N*-phthalimido)glutarimide.

It was withdrawn from the market as its use during pregnancy was associated with malformations of the fetus (phocomelia or shortening of the members).

In 1984, it is published that this tragedy would not have occurred if, instead of having used the racemic, the enantiomer (*R*) had been launched to the market, because in experimental conditions only the enantiomer (*S*)-(–) exerted toxic and teratogenic effects on the embryo, after intraperitoneal administration.

This statement is wrong for two reasons:

(1)   It is based on low-confidence biological data since experimental studies were performed on mice, a species that is considered insensitive, and at very high doses; however, later works in rabbits (species that is the most sensitive to thalidomide) showed the same teratogenic potency of both enantiomers.

(2) The stereocenter of thalidomide is unstable in protonated media and suffers a rapid reversal of its configuration, so both enantiomers are racemized quickly and degraded by opening the ring of the glutarimide moiety, a process that takes place faster in vivo than in vitro. Therefore, if there were differences of toxicity between both enantiomers, their rapid racemizations in vivo would have prevented the commercialization of the nonpoisonous enantiomer.

This case shows the importance of considering the data as a whole and of avoiding to draw quick conclusions, as tempting as they are.

1.6 Origin of drugs

- Drugs of natural origin:
  1. Vegetal (25%): there are many extractable drugs from phanerogams, and among them, we highlight the alkaloids, such as quinine, reserpine, and morphine.
  2. Animal (18%): e.g. insulin, a hypoglycemic substance obtained from ox fresh pancreas, sex hormones, and corticosteroids.
  3. Mineral (7%): this is the case of aluminum salts to alleviate acidity of the stomach, or talc (silicates) to relieve pruritus and as a base for ointments.
- Drugs of semisynthetic origin: they are obtained by partial synthesis from a structure of natural origin, for example, semisynthetic penicillins.
- Drugs of synthetic origin: estradiol is a natural estrogen, ethinylestradiol is a semisynthetic derivative, while diethylstilbestrol is a synthetic product (Fig. 1.5).

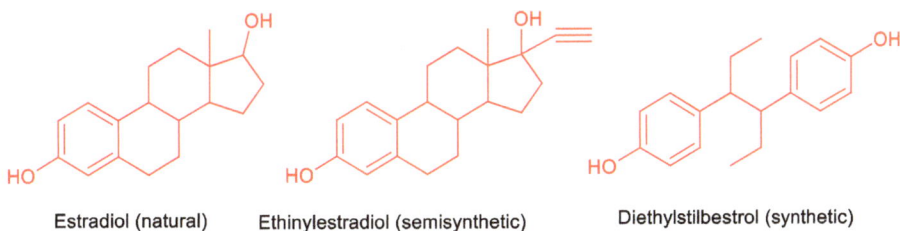

Estradiol (natural)   Ethinylestradiol (semisynthetic)   Diethylstilbestrol (synthetic)

Fig. 1.5: Examples of natural, semisynthetic, and synthetic drugs.

These last two groups constitute ≈ 50% and are those that are usually studied by PC.

1.7 Other definitions

– The action of a drug refers to the modification that it produces in the functions of the organism, in the sense of increasing or diminishing them. Drugs never create new functions or alter the characteristics of the system on which they act: they only modify them. The effect or response of a drug is the manifestation of pharmacological action, which can be appreciated by the observer's senses or by simple devices (Fig. 1.6).

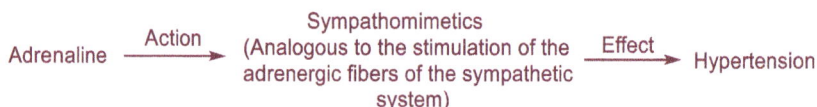

Adrenaline —— Action ——→ Sympathomimetics (Analogous to the stimulation of the adrenergic fibers of the sympathetic system) —— Effect ——→ Hypertension

Fig. 1.6: Action and effect of adrenaline (also called epinephrine).

– Molecules that show biological activity are called hits.

1.8 Classification of drugs

Pharmacological effect: Drugs are classified depending on the biological effect they produce, e.g. analgesics. However, there are many biological goals and mechanisms by which analgesics can have an analgesic effect. Therefore, it is not possible to identify a common characteristic shared by all analgesics. For example, Aspirin® and morphine act on different goals and have no structural relationship.

Chemical structure: Many drugs have a common skeleton and are grouped according to this criterion, e.g. penicillins, barbiturates, opiates, and catecholamine. For example, penicillins contain the β-lactam ring and kill the bacteria by the same mechanism. However, it is not foolproof. Sulfonamides have a common structure and are fundamentally antibacterial agents. However, some sulfonamides are used for the treatment of diabetes.

Biochemical process: Generally, a chemical messenger or neurotransmitter, e.g. antihistamines and cholinergics. It is more specific than that of the pharmacological effect, since it identifies the system on which the drugs act.

Molecular objective: There are compounds that are grouped according to the enzyme or receptor with which they interact; e.g. anticholinesterases are a group of drugs that act by inhibiting the enzyme acetylcholinesterase. This is the most useful classification with respect to PC.

1.9 Process of discovery of a drug

Successful research and development of a drug carries with it gigantic costs that can fluctuate around 1.4 billion dollars (Fig. 1.7). These astronomical costs are due to the fact that of 5,000 candidate molecules only 1,000 become objective compounds and a dozen become "leaders", that is to say, they are promising, and of those thousands only one will arrive without problems up to the last phase of development. However, if only one of these drugs complies with all basic and clinical evaluations and is patented, the initial multimillion-dollar investment will be virtually paid off.

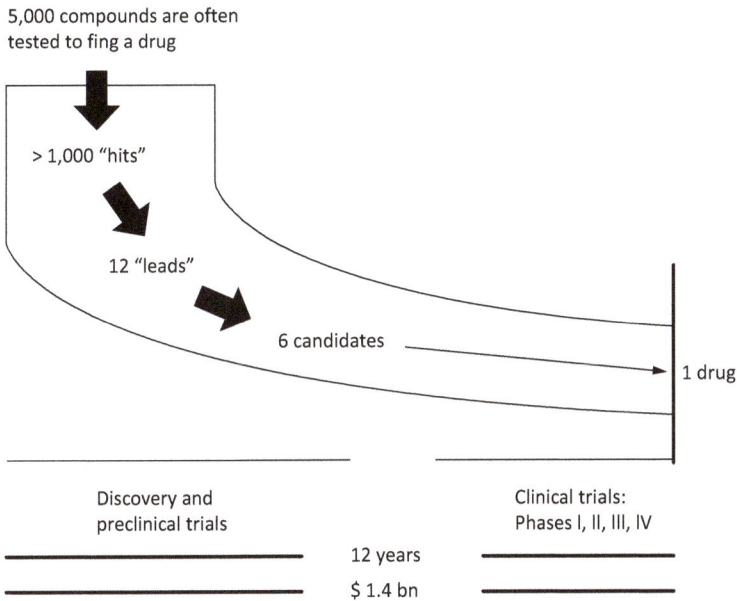

5,000 compounds are often tested to fing a drug

> 1,000 "hits"

12 "leads"

6 candidates

1 drug

Discovery and preclinical trials

Clinical trials: Phases I, II, III, IV

12 years

$ 1.4 bn

Fig. 1.7: "Bottleneck" process outlining the difficulty of the drug discovery process.

A drug may take 12 years from the initial discovery state to reach the market, and while estimates of costs vary, the Association of the British Pharmaceutical Industry puts it at $1.4 billion per drug. Just 1 in 5,000 drug candidates make it all the way from the drug discovery phase to licensing approval. If the pharmaceutical company patents compounds in the early stages, it has only 5–6 years to recoup the heavy investment and/or make a profit.

The number of biological targets is expected to increase tenfold for the treatment of diseases with the increasing advancement of proteomics, genomics, and molecular pharmacological techniques. Also being worked on is the development of other new drugs, including gene therapy and nanotechnology. The goal of PC is to make the drug development process more cost-effective, shortening the time between discovery, preclinical testing, and registration steps. This is why the rational design of drugs is at present of such a large dimension.

1.10 Phases of clinical studies

Phase I: It is not yet useful to test the efficacy of the drug. Phase I studies attempts to determine the potential toxic effect of the drug in humans by administering small doses to a small number of healthy volunteers. In addition, pharmacokinetic and pharmacodynamic parameters are measured.

Phase II: In Phase II studies, the drug is administered to a small number (greater than in Phase I) of people suffering from a particular disease in order to clearly demonstrate the potential therapeutic benefit of the drug. This is a first measure of efficacy, but above all it is to determine the best doses and modes of distribution (oral, intravenous, intramuscular, etc.) and to confirm the results of the Phase I tests.

Phase III: Phase III studies involve a larger number of patients with an established dose range and final administration form in order to redefine the knowledge obtained in Phase II studies. It aims to confirm the efficacy and safety of the drug on a large scale. Once Phase III studies are completed, the pharmaceutical company can apply for approval of the drug for marketing. This research process can last from 7 to 15 years and is the most expensive of the entire research process.

Phase IV: An even larger population or a specific subgroup, evaluation of the long-term effects of the drug, or even its test in other indications.

1.11 Key notes: introduction

Pharmaceutical chemistry: PC deals with the design and synthesis of novel drugs, based on an understanding of how they work at the molecular level. A useful drug must interact with a molecular target in the body (pharmacodynamics) and also be capable of reaching the target (pharmacokinetics).

Drugs: Drugs are normally low-molecular-weight chemicals that interact with macromolecular targets in the body to produce a pharmacological effect. This effect may be beneficial or harmful depending on the drug used and the dose administered.

Classification of drugs: Drugs can be classified according to their pharmacological effect, the particular biochemical process they affect, the type of structures they have, or the molecular target with which they interact. The last classification is the most useful one in PC.

1.12 Key notes: clinical trials

Definition: Clinical trials are carried out to test the therapeutic effects of new drugs and to ensure that they have no unacceptable side effects. There are four phases:

Phase I: Phase I trials are normally carried out on *healthy volunteers* to establish dosing levels and to carry out pharmacokinetic studies. The therapeutic effect is not tested.

Phase II: Phase II studies are carried out on patients for a particular indication. One group of patients receives the drug and another one receives a placebo or a conventional drug. Neither patient nor doctor knows which patient receives placebo or drug. Different dosage levels and regimes are used on different groups to establish the better dosage regimen. These studies demonstrate whether the drug is therapeutically useful and whether it has any side effects.

Phase III: Phase III studies are carried out in a similar fashion to Phase II studies, but on a larger number of patients in order to get statistical proofs, both of the drug efficacy and its safety.

Phase IV: Phase IV studies continue after the drug has gone onto the market. They are designed to study the effects of long-term use and to identify any rare side effects that may arise.

2 Drug nomenclature

2.1 Goals

- Knowledge of drug nomenclature: types
- Knowledge of the International Nonproprietary Name
- Knowledge of the systematic nomenclature: the IUPAC rules for naming organic molecules
- Knowledge of other nomenclatures

2.2 Introduction

Nowadays, knowledge of chemical nomenclature is indispensable; however, most texts are limited to the handling of the nomenclature of isolated functional groups, and no integration is achieved in the application of the rules of nomenclature for complex structures. It is intended that by the end of this book, readers will be able to establish the name of a compound from its structure and vice versa, regardless of its complexity. Important changes in the recommendations affecting the nomenclature of organic compounds appear in the book *Nomenclature of Organic Chemistry: IUPAC Recommendations and Preferred Names 2013*.

2.3 Nomenclature of drugs

Drugs can be named in several ways:

Coded name: Usually with the initials of the laboratory, chemist, or research team who prepared or tested the drug for the first time, followed by a number. This name does not tell us anything about the structure or the pharmacological action.

Chemical name: Describes in an unambiguous way the structure of the drug. It is produced in accordance with IUPAC standards. Since the chemical name can be very complicated, it is not suitable for routine use.

Registered name: The name given by each manufacturer. Normally, a drug is patented by different industries, and therefore it can appear on the market with several names. It is symbolized with the symbol ® to the right and top of the name. The first letter of each word that is part of the name must be capitalized. It says nothing about structure or action.

International nonproprietary name (INN): This is the name of the drug with which it is identified as a specific substance and independent of its manufacturer. The first

https://doi.org/10.1515/9783111316901-002

letter must be lowercase or all uppercase. Since 1976, the World Health Organization is responsible for developing international standards. The name should be brief, concise, and meaningful. The relationship between substances of the same group and with the same pharmacological activity is evidenced by the use of some characteristic particle. Table 2.1 shows some of these particles:

Tab. 2.1: Particles used in the construction of the international common denominations (selection).

Particle	Category	Compound
-azepam	Benzodiazepines	Diazepam
-bamate	Diol anxiolytics	Meprobamate
-barb-	Barbiturates	Phenobarbital
-caine	Local anesthetics	Procaine
ceph-	Cephalosporins	Cefalotine
-cillin	Penicillins	Ampicillin
sulfa-	Sulfonamides	Sulfathiazole

Acronyms: These are used in drugs of Anglo-Saxon countries (Fig. 2.1).

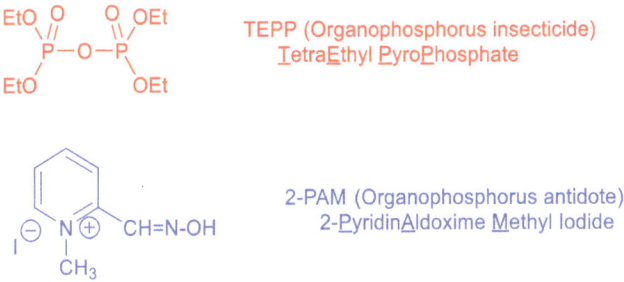

TEPP (Organophosphorus insecticide)
TetraEthyl PyroPhosphate

2-PAM (Organophosphorus antidote)
2-PyridinAldoxime Methyl Iodide

Fig. 2.1: Acronyms used to name some drugs.

2.4 Systematic chemical names

Table 2.2 shows the order of various chemical-organic functions, from the major importance (ammonium salts) to the minor one (ethers).

Tab. 2.2: Main functional groups in descending order of priority.

Generic name	Functional group	Prefix (substituent)	Suffix (principal chain)
1. Ammonium salts	$R_4N^+X^-$		Ammonium halides
2. Carboxylic acids	R-COOH	Carboxy-	Carboxylic acid -oic/-ic acid
3. Anhydrides	$(R-CO)_2O$		-oic anhydride
4. Esters	R-COOR'	R-Oyloxy- R-Carbonyloxy- R'-Oxycarbonyl-	R'-Carboxylate R'-ate
5. Acid halides	R-CO-X	Haloformyl-	-oyl halide
6. Amides	R-CONH-R'	R'-Carbamoyl- R-Amido- R-Carboxamido-	R-Amide
7. Nitriles	R-CN	Cyano-	-carbonitrile -nitrile
8. Aldehydes	R-CHO	Formyl-oxo-	-carbaldehyde -al
9. Ketones	-CO-	Oxo-	-one
10. Alcohols, phenols	R-OH, Ar-OH	Hydroxy-	-ol
11. Thiols	R-SH	Mercapto-	-thiol
12. Amines	$R-NH_2$	Amino-	-amine
13. Ethers	R-O-R'	Oxy-	-ether

2.5 Polyfunctional acyclic compounds

First, the main function has to be identified. Next, a structural fragment has to be chosen, and it will be considered as main, applying the norms of the IUPAC.

The hypnoanalgesic methadone contains two functions (ketone and amine), of which ketone is the priority one and therefore named as a suffix. This numbers the end closest to the main function and chooses as the main chain, which is the longest containing the main function. An example is shown in Fig. 2.2.

The drug primocarcin has a secondary ketone function (prefix) and two amide (priority) functions. In the ester and amide functions, the chain to be considered is that which contains the carbonyl group. Since it is not possible to find a chain containing two amide functions, the main chain will include a chain of six carbons and a double bond (Fig. 2.3).

In the hypnotic chlorhexadol, there is no carbon chain containing the two alcohol functions, with priority over the ether function (halogen is not considered a functional

Principal function

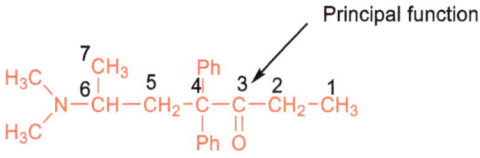

6-Dimethylamino-4,4-diphenylheptan-3-one

Fig. 2.2: Methadone (hypnoanalgesic).

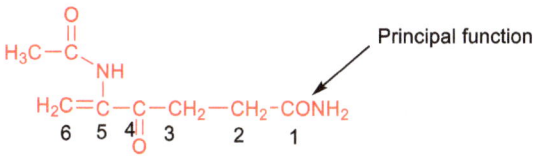

Principal function

5-Acetamido-4-oxohex-5-enamide

Fig. 2.3: Primocarcin (antineoplastic and antimicrobial agent).

group). The chain containing five carbons is selected as the principal one. It is numbered by giving the lowest number to the main function (Fig. 2.4).

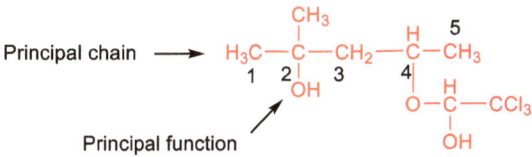

Principal chain

Principal function

2-Methyl-4-(2,2,2-trichloro-1-hydroxyethoxy)pentan-2-ol

Fig. 2.4: Chlorhexadol (hypnotic agent).

Finally, let us look at the case of the acetylcholine cholinergic neurotransmitter (Fig. 2.5).

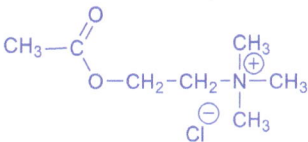

Fig. 2.5: Acetylcholine (AcC).

The main function is the ammonium group and hence the name is (2-acetoxyethyl)trimethylammonium chloride.

2.6 Monocyclic compounds

The same criteria are used in choosing the main chain or main cycle. Consider the following difunctionalized molecule (Fig. 2.6).

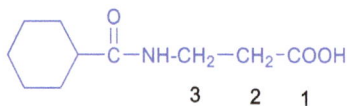

Fig. 2.6: A difunctionalized molecule.

The main function is the carboxylic acid, and the amide function is the secondary one. The correct name is 3-(cyclohexanecarboxamido)propanoic acid.

Procaine is an anesthetic agent indicated for production of local anesthesia, particularly for oral surgery. It is named as a benzoic acid derivative, since the carbonyl group of the ester function is attached directly to the benzene ring. There are three different ways of naming procaine (Fig. 2.7).

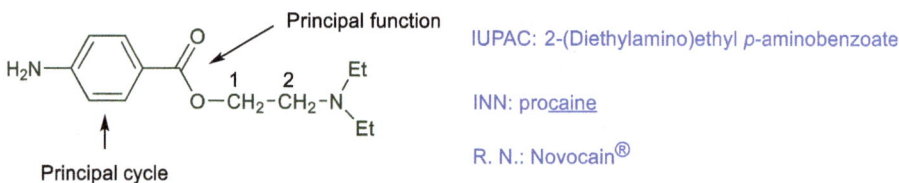

IUPAC: 2-(Diethylamino)ethyl p-aminobenzoate

INN: procaine

R. N.: Novocain®

Fig. 2.7: Several ways of naming procaine.

In adrenaline (neurotransmitter), the main chain is the one containing the alcoholic hydroxyl group (preferred over phenolic hydroxyls). In addition, a correct nomenclature must include the configuration of the stereogenic centers when indicated in the structure (Fig. 2.8).

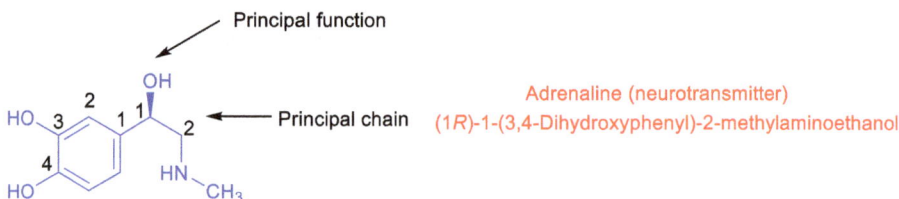

Adrenaline (neurotransmitter)
(1R)-1-(3,4-Dihydroxyphenyl)-2-methylaminoethanol

Fig. 2.8: Adrenaline (neurotransmitter).

Another example is the antimicrobial chloramphenicol, for which two systematic names are given (**A** and **B**). If the carbonyl group corresponding to the major function (carboxamide) is chosen as the main chain, a complex systematic name (**A**) results.

For the type of RCONHR′ amides, where R′ is much more complex than R, the IUPAC rules support alternative **B**, where R′ is considered prime (Fig. 2.9).

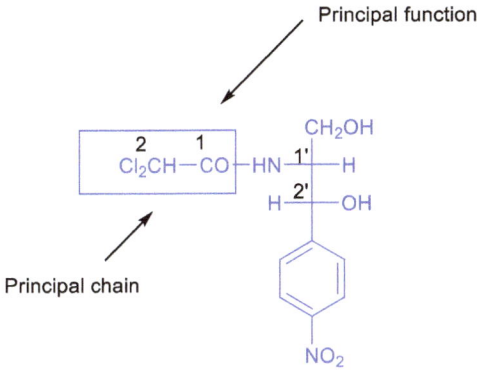

A. **2,2-Dichloro-*N*-[(1*R*,2*R*)-1,3-dihydroxy-1-(4-nitrophenyl)propan-2-yl]acetamide**

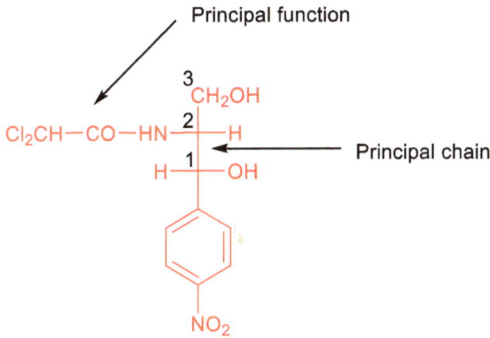

B. (1*R*,2*R*)-2-Dichloroacetamido)-1-(4-nitrophenyl)propan-1,3-diol

Fig. 2.9: Chloramphenicol (antibacterial agent).

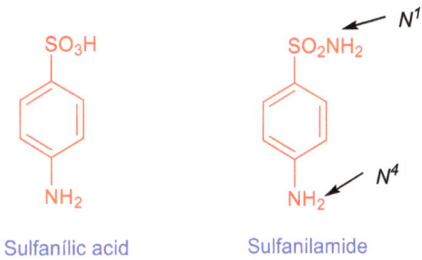

Sulfanílic acid   Sulfanilamide

Fig. 2.10: Sulfanilic acid and sulfanilamide.

For the nomenclature of antibacterial agents, sulfonamides, the common name sulfanil-amide for the 4-aminobenzenesulfonamide unit (from sulfanilic acid) is taken as the basis. Prefixes N^1 and N^4 are used as locators of their two nitrogen atoms (Fig. 2.10).

Example: Sulfanitran (Fig. 2.11).

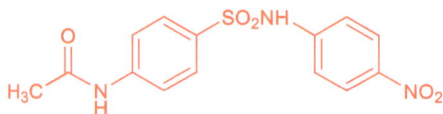

N^4-Acetyl-N^1-(p-nitrophenyl)sulfanilamide

Alternatively named as:

N-{4-[N-(4-Nitrophenyl)sulfamoyl]phenyl}acetamide

Fig. 2.11: Sulfanitran (antibacterial agent).

2.7 Polycyclic aromatic hydrocarbons

Polycyclic aromatic hydrocarbons are organic compounds formed by two or more fused aromatic rings. The rings may be in a straight, angled, or clustered shape. The simplest condensed structure formed by only two aromatic rings is naphthalene, and the three-ring compounds are anthracene and phenanthrene.

The prefix for this hydrocarbon corresponds to the polycyclic system with the maximum number of nonaccumulated double bonds. If this number is not reached, the letter H with a number will indicate the state of hydrogenation. Most of the condensed polycyclic hydrocarbons have a common name (Fig. 2.12).

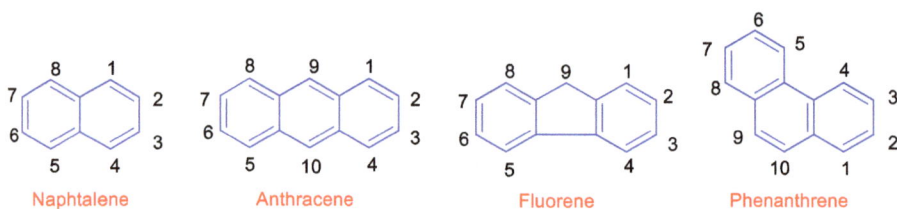

Fig. 2.12: Some condensed polycyclic hydrocarbons and numbering.

Compounds without a common name are denominated by combining the fundamental component (common name system with the highest number of rings) with the prefixes representing the remaining cyclic components fused with it. The names of hydrocarbons themselves are used as prefixes; some of them abbreviated benzo, naphtho, phenanthro, and anthra.

The face by which the cyclic component is attached to the fundamental one is designated by a locator (a cursive letter *a*, *b*, *c*, . . .) which is ascribed in alphabetical order on each side of the fundamental nucleus, beginning with the carbon atoms 1 and 2 (side *a*).

If the condensed system has two rings and does not have a common name, the fundamental component will be the major ring named as cycloheptene, cyclooctene, cyclononene, etc., which indicates the maximum number of unsaturations (Fig. 2.13).

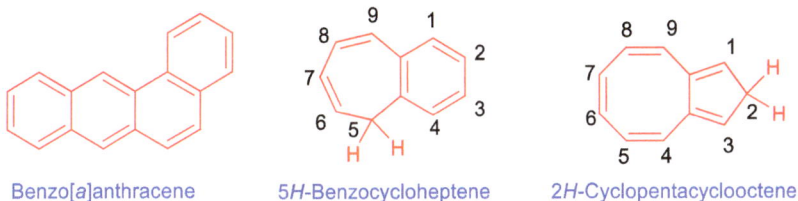

Benzo[*a*]anthracene   5*H*-Benzocycloheptene   2*H*-Cyclopentacyclooctene

Fig. 2.13: Some condensed bicyclic hydrocarbons, in which one of the rings does not have a common name.

A condensed polycyclic system must be numbered for the location of the substituents or functional groups. First, the polycyclic system must be oriented, to finally proceed to the numbering of each vertex. To orient the polycyclic system, it must be drawn so that
(a) a maximum number of rings are aligned in a horizontal row;
(b) a maximum number of rings is located in the first upper right quadrant; the vertical line is drawn on the inner side of the leftmost cycle;
(c) when two orientations meet the above requirements, the one with the lowest number of rings below the horizontal and to the left is chosen.

An example is the orientation of chrysene (Fig. 2.14).

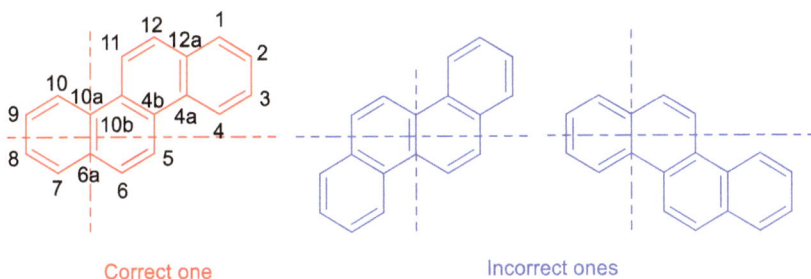

Correct one   Incorrect ones

Fig. 2.14: Orientation of chrysene.

Once the system has been oriented, number 1 is assigned to the position immediate to the condensation in the ring that is (a) higher and (b) more to the right-hand side. The numbering is continued clockwise. The carbon atoms common to two or more rings are not numbered correlatively, but are designated by the addition of the letters *a*, *b*, *c*, and so on, to the immediately preceding numeral. The inner atoms of the polycyclic system follow the last exterior, also clockwise. An example is fluorene (Fig. 2.15).

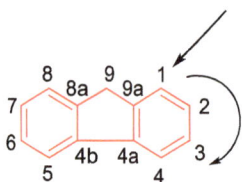

Fig. 2.15: Numbering of fluorene.

2.8 Partly or fully hydrogenated polycyclic hydrocarbons

If the condensed hydrocarbon leaves a hydrogen atom free in one position, it is named as *H*. If several positions are left free, they are named dihydro-, tetrahydro-, and hexahydro-, preceded by corresponding locators (even multipliers). If the molecule does not contain any double bonds, the prefix perhydro- is used. If there is an odd number of hydrogenated positions, a hydrogen with the lowest possible locator and the other even-numbered positions are indicated as explained (Fig. 2.16).

Tetracyclines are an important group of antibiotics structurally derived from naphthacene, partially hydrogenated, and with a high degree of functionalization. In almost

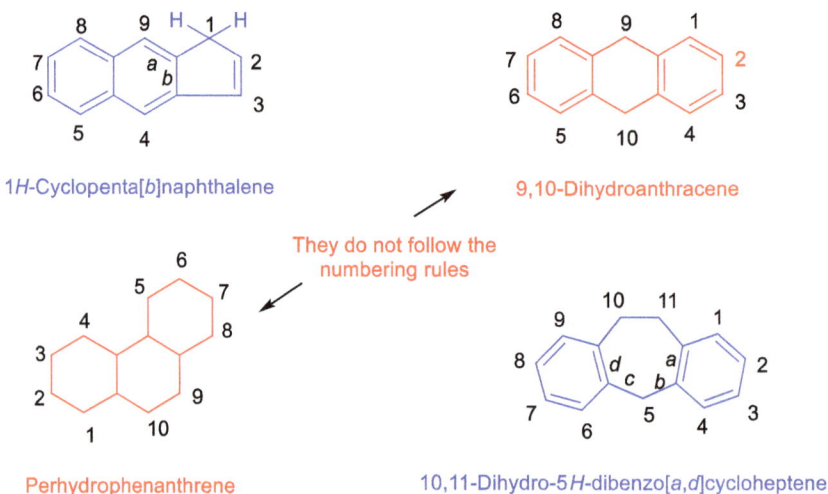

1*H*-Cyclopenta[*b*]naphthalene

9,10-Dihydroanthracene

They do not follow the numbering rules

Perhydrophenanthrene

10,11-Dihydro-5*H*-dibenzo[*a*,*d*]cycloheptene

Fig. 2.16: Partly or fully hydrogenated polycyclic hydrocarbons.

all cases (all natural and many of the semisynthetic ones), the main function is a carbox-amide, whose preference in numbering gives it position 2. Thus, the tetracyclines will be naphthacene-2-carboxamides, although they are usually drawn oriented with position 1 in the lower right side. An example would be chlortetracycline (Aureomycin®, Fig. 2.17).

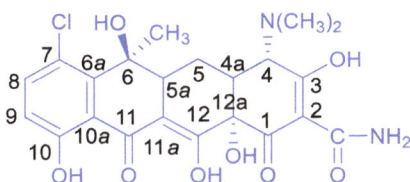

Fig. 2.17: Chlortetracycline (Aureomycin®).

The hydrogenated positions are eight: 1, 4, 4a, 5, 5a, 6, 11, and 12a. The complete systematic name would be as follows: 7-chloro-4-dimethylamino-3,6,10,12,12a-pentahydroxy-6-methyl-1,11-dioxo-1,4,4a,5,5a,6,11,12a-octahydronaphthacene-2-carboxamide. Note that positions 1 and 11 are referred to as hydrogenated, in spite of having carbonyl groups, since these ketone functions are prefixed "oxo" in the name and do not give rise to the suffix "one".

2.9 Bicyclic hydrocarbons

2.9.1 Bridge systems

They are compounds that share two atoms called bridgeheads. They are named by adding the prefix "bicyclo" followed by the number of atoms between the bridgeheads (in squared brackets), separated by points and ordered from major to minor, ending with the name of the alkane resulting from the sum of all the atoms of carbon of the bicyclic base system (Fig. 2.18).

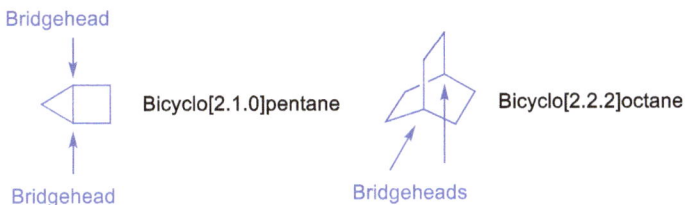

Fig. 2.18: Bridge systems.

The numbering of these systems is started at the bridgehead, continued by the longest chain to the other bridgehead, then the intermediate chain, and finally, the shortest one (Fig. 2.19).

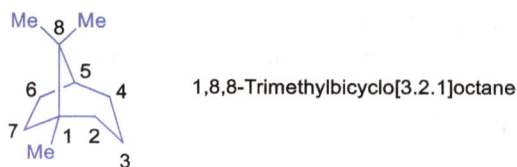

1,8,8-Trimethylbicyclo[3.2.1]octane

Fig. 2.19: Numbering of a bridge system.

2.9.2 Bicyclic systems with heteroatoms

The type and position of heteroatoms (atoms other than carbon and hydrogen) are indicated, respectively, by a prefix and a numbering. The three most frequent heteroatoms, *O*, *N*, and *S*, are indicated by the prefixes "oxa", "aza", and "thia", respectively. The order of priority between them is *O* > *S* > *N*. Consider two examples of pharmaceutical significance, beginning with a tropane alkaloid:

(a) **Atropine (antimuscarinic drug).** The compounds of this family contain the fundamental nucleus of 8-methyl-8-azabicyclo[3.2.1]octane, commonly known as "tropane" (Fig. 2.20).

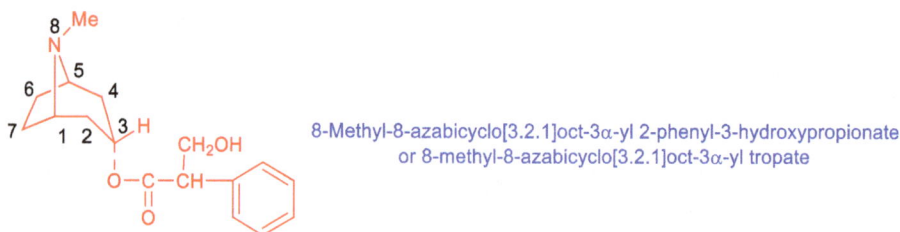

8-Methyl-8-azabicyclo[3.2.1]oct-3α-yl 2-phenyl-3-hydroxypropionate or 8-methyl-8-azabicyclo[3.2.1]oct-3α-yl tropate

Fig. 2.20: Atropine.

(b) We have another related example in the local anesthetic cocaine (Fig. 2.21).

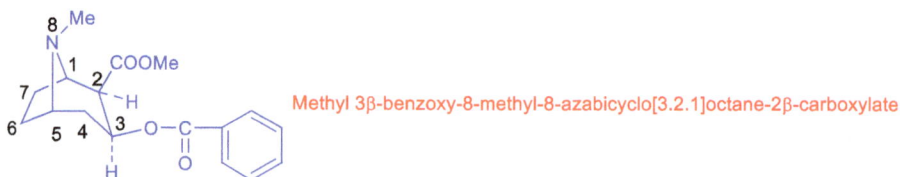

Methyl 3β-benzoxy-8-methyl-8-azabicyclo[3.2.1]octane-2β-carboxylate

Fig. 2.21: Cocaine.

(c) **Penicillins and cephalosporins** are included within the general class of "β-lactams" because they have a four-membered lactam as a common structural feature. In all cases, this ring is condensed with another five- or six-membered heterocycle, giving rise to penicillins and cephalosporins, respectively. We then establish the systematic names of penicillin G and cephalothin (Fig. 2.22).

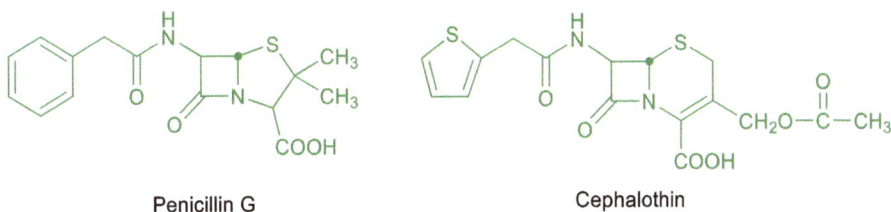

Penicillin G   Cephalothin

Fig. 2.22: A penicillin and a cephalosporin.

For the nomenclature of these drugs, there are several possibilities, from a totally systematic name, through semisystematic names, to a common serial name.

Let us look at the different forms of nomenclature of the three nuclei (Fig. 2.23).

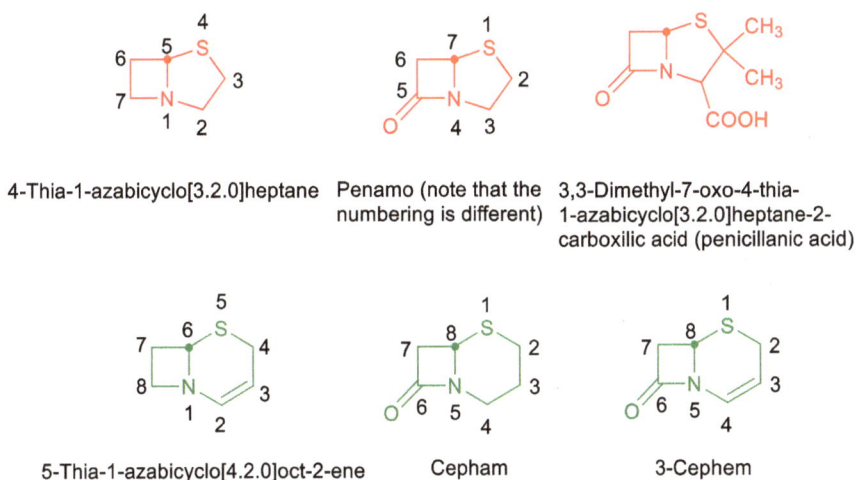

4-Thia-1-azabicyclo[3.2.0]heptane   Penamo (note that the   3,3-Dimethyl-7-oxo-4-thia-
                                      numbering is different)   1-azabicyclo[3.2.0]heptane-2-
                                                                carboxilic acid (penicillanic acid)

5-Thia-1-azabicyclo[4.2.0]oct-2-ene   Cepham   3-Cephem

Fig. 2.23: Several possibilities of naming penicillins and cephalosporins.

Consequently, the systematic names of penicillin G and cephalothin are as follows:

Penicillin G: 6-(Phenylacetamido)-3,3-dimethyl-7-oxo-4-thia-1-azabicyclo[3.2.0]-heptane -2-carboxylic acid; alternative: 6-(phenylacetamido)penicillanic acid.

Cephalothin: 3-(Acetoxymethyl)-7-[2-(2-thienyl)acetamido]-8-oxo-5-thia-1-azabicyclo [4.2.0]oct-2-ene-2-carboxylic acid; alternative: 7-(2-(2-thienyl)acetamido] cephalosporanic acid. It is convenient to show the names of two radicals derived from thio-phene (Fig. 2.24).

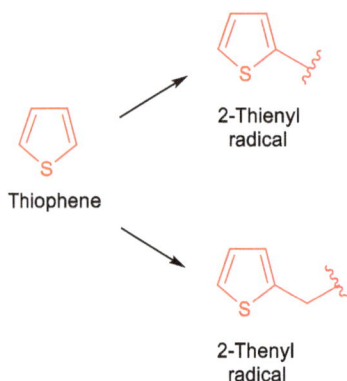

2-Thienyl
radical

Thiophene

2-Thenyl
radical

Fig. 2.24: Two radicals derived from thiophene.

2.9.3 Spiro or spirane systems

It is said that two rings are united to form a spiro structure when they share a single atom, and this is the unique union that exists between them. The common atom is called the "spiro atom". To name them, there are two valid options. In the first, spiro is written first, followed by square brackets of the number of carbon atoms attached to the spiro atom, ordered from smaller to larger rings, separated by a point and then the linear hydrocarbon containing the same total number of carbon atoms is named. Let us look at two examples (Fig. 2.25).

Spiro carbon atom

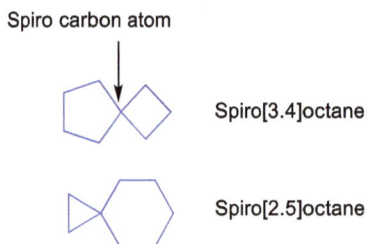

Spiro[3.4]octane

Spiro[2.5]octane

Fig. 2.25: Examples of spiro compounds.

They are numbered beginning with the carbon atom of the smallest chain next to the spiro carbon, continue through the spiro carbon, and finally continue to number the major cycle. If there are substituents, try to give them the lowest possible locant (Fig. 2.26).

The second or alternative option to name these spiro compounds is to put the name of the major cycle before the smaller one and insert the word "spiro". Each cycle maintains its own numbering, although the carbons of the mentioned cycle are marked with primes at the end (Fig. 2.27).

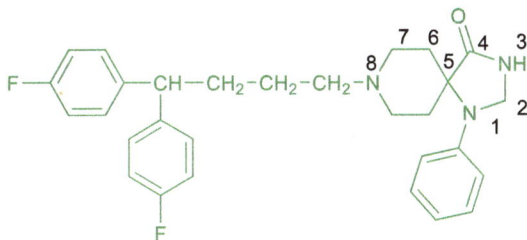

1-Phenyl-8-[4,4-bis(p-fluorophenyl)butyl]-1,3,8-triazaspiro[4.5]decan-4-one

Fig. 2.26: Fluspirilene (antipsychotic agent).

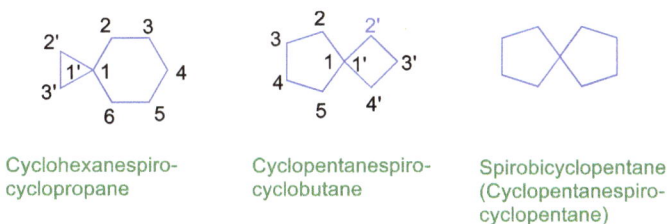

Cyclohexanespiro-
cyclopropane

Cyclopentanespiro-
cyclobutane

Spirobicyclopentane
(Cyclopentanespiro-
cyclopentane)

Fig. 2.27: Another form of naming and numbering spiro compounds.

2.10 Heterocycles

Heterocycle denominates any cyclic system in which one or more links are constituted by heteroatoms, i.e. atoms other than carbon. Most heterocyclic systems have a common name: pyrrole, furan, thiophene, pyridine, indole, and quinoline. Those that do not have a common name are named combining a series of prefixes and roots. The prefixes indicate the heteroatom: oxa- (O), thia- (S), aza (N), in this order of priority, if there is more than one heteroatom in a cycle. The roots indicate the size of the ring and its degree of saturation. Hantzsch and Widman introduced a nomenclature system for five- and six-membered nitrogen heterocycles which has spread to other heterocyclic rings.

The Hantzsch–Widman method is the most often used to name heterocycles that do not have bridges. The roots corresponding to heterocycles of three to ten atoms are indicated in Tab. 2.3. It consists of five sections: the first column shows the size of the cycle; the following two contain the roots to be used when the heterocycle contains nitrogen and the latter two the roots for heterocycles without nitrogen. In addition, the particles are different if the cycle contains the maximum number of nonaccumulated double bonds (unsaturated, columns 2 and 4) or fully saturated (columns 3 and 5). Nitrogen heterocycles of six to ten members are an exception, since their total saturation is expressed in the usual way with the prefix "perhydro" preceding the name of the unsaturated compound.

Tab. 2.3: Common name endings for heterocyclic compounds.

Size	Nitrogen cycles		Non-nitrogen cycles	
	Unsaturated	**Saturated**	**Unsaturated**	**Saturated**
3	-irine	-iridine	-irene	-irane
4	-ete	-etidine	-ete	-etane
5	-ole	-olidine	-ole	-olane
6	-ine	*	-in	-ane
7	-epine	*	-epin	-epane
8	-ocine	*	-ocin	-ocane
9	-nonine	*	-nonin	-nonane

*Perhydro followed by the termination of the unsaturated heterocycles.

In those heterocycles in which two or more identical heteroatoms exist, they will be indicated by a multiplier di, tri, etc., before the corresponding prefix. The numbering of cycles with a single heteroatom always begins in this one. If the heteroatoms are of a different nature, they are numbered beginning with the higher priority O > S > N. Before developing the nomenclature of some examples of heterocyclic drugs, we will proceed to name the carbonic acid derivatives, which are important reagents in the synthesis of drugs or form part of the structures of important groups of drugs (Fig. 2.28).

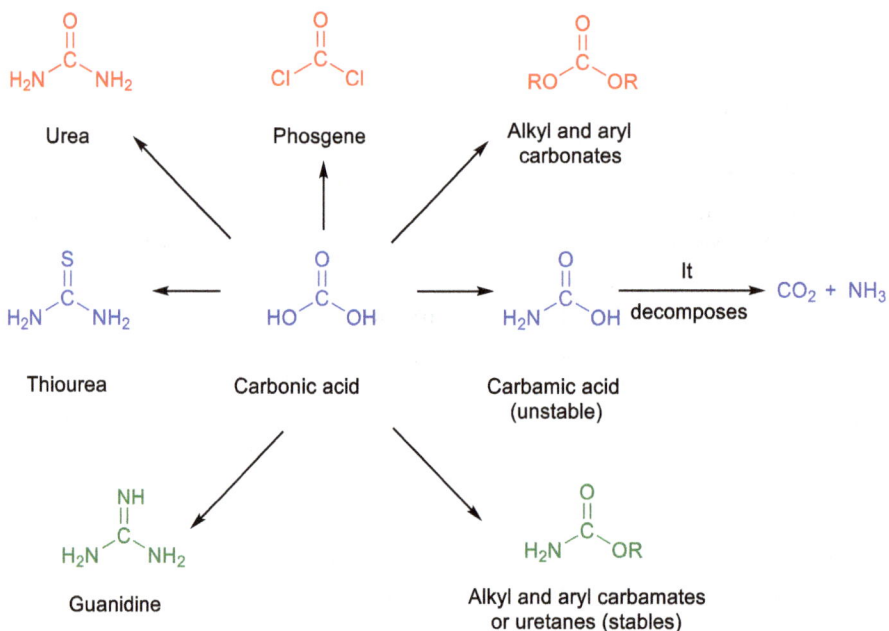

Fig. 2.28: Carbonic acid derivatives.

Consider the antihypertensive agent, guanethidine; the prefix perhydro will be used to indicate the saturation (Fig. 2.29).

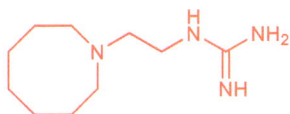

N-[(2-Perhydro-1-azocine-1-yl)ethyl]guanidine

Fig. 2.29: Guanethidine (antihypertensive agent).

Consider the antiulcer agent cimetidine (Fig. 2.30).

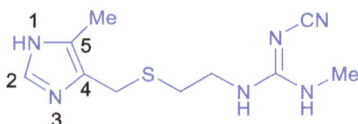

2-Cyano-3-methyl-1-{[2-(5-methyl-1H-imidazole-4-yl)methylthio]ethyl}guanidine

Fig. 2.30: Cimetidine (antiulcer agent).

The antiulcer drug ranitidine contains three amino groups, two of which are linked to the same carbon of a unit of ethylene. Accordingly, the fundamental structural unit is a vinylidenediamine, or 1,1-ethylenediamine (Fig. 2.31).

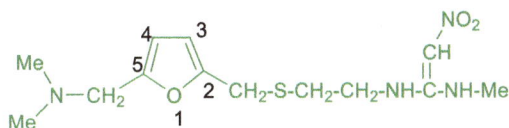

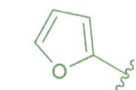

2-Furyl radical   2-Furfuryl radical

The divalent radicals $CH_2=C=$, $-CH=CH-$ and $-CH_2CH_2-$ are denominated as vinylidene, vinylene and ethylene, respectively

N-{2-[(5-Dimethylaminomethylfurfuryl]thioethyl}-N'-methyl-2-nitrovinylidenediamine (or -1,1-ethylenediamine)

Fig. 2.31: Ranitidine (antiulcer agent).

A group of great importance within the heterocyclic nucleus drugs are the so-called benzodiazepines, with multiple actions but mainly anxiolytic, anticonvulsive, sedative, and hypnotic drugs. One of the simplest drug is diazepam (Fig. 2.32).

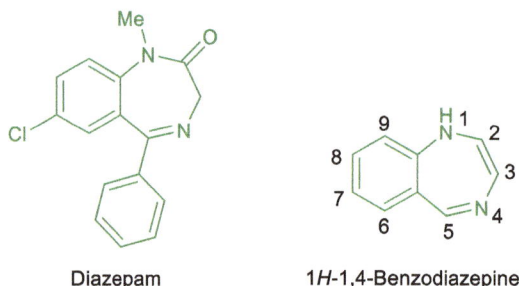

Diazepam   1*H*-1,4-Benzodiazepine

7-Chloro-5-phenyl-1-methyl-1,3-dihydro-1,4-benzodiazepine-2-one

Fig. 2.32: Diazepam.

In *Chemical Abstracts*, diazepam receives a name almost equivalent, although with the incorporation of hydrogen indicated by "2*H*", which the nomenclature IUPAC considers unnecessary.

Another example is chlordiazepoxide or Librium® (Fig. 2.33).

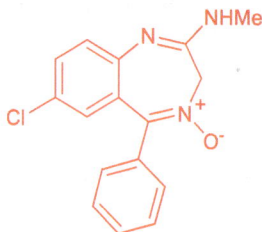

7-Chloro-5-phenyl-2-methylamino-3*H*-1,4-benzodiazepine-4-oxide

Fig. 2.33: Librium®.

The frequency of heterocyclic systems in compounds of natural origin determines the proliferation of common names, among which we refer to the most frequent ones in Fig. 2.34 (five-membered heterocycles) and Fig. 2.35 (six-membered heterocycles):

2.10.1 Condensed heterocycles

Most important is to choose a fundamental component and name the secondary components as prefixes, usually ending in "o". Where the name of a (isolated) component

	Azole **Pyrrole**		Azolidine **Pyrrolidine**
	Oxol **Furan**		Oxolane **Tetrahydrofuran**
	Thiol **Thiophene**		Thiolane **Tetrahydrothiophene**
	1,3-Diazole **Imidazole**		1,3-Diazolidine **Imidazolidine**
	1,2-Diazole **Pyrazole**		1,2-Diazolidine **Pyrazolidine**
	1,3-Oxazole **Oxazole**		1,2-Oxazole **Isoxazole**
	1,3-Thiazole **Thiazole**		1,2-Thiazole **Isothiazole**

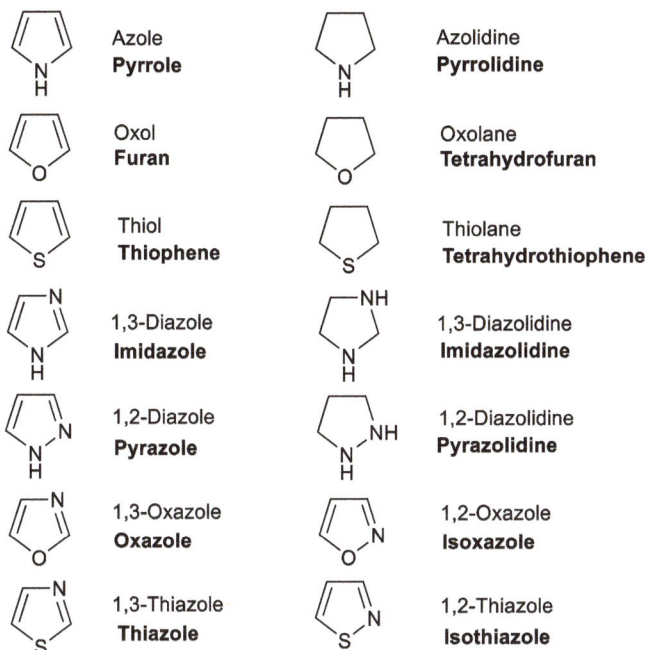

Fig. 2.34: Selected five-membered heterocyclic skeletons.

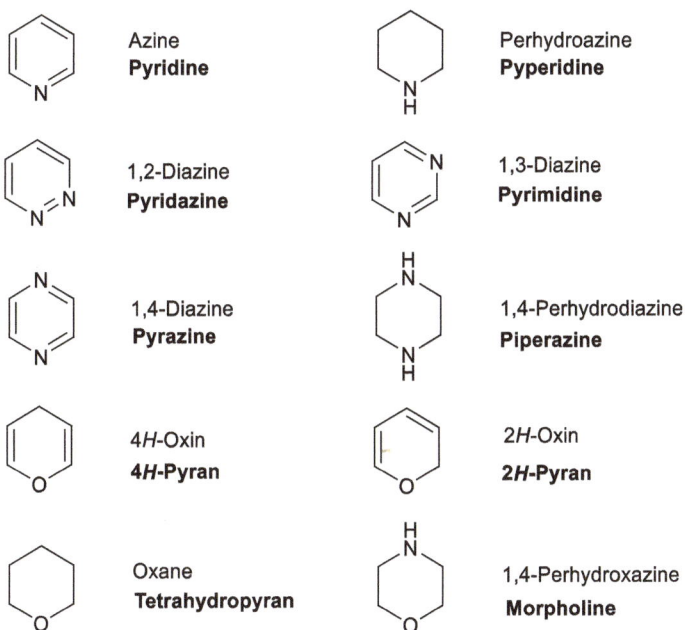

	Azine **Pyridine**		Perhydroazine **Pyperidine**
	1,2-Diazine **Pyridazine**		1,3-Diazine **Pyrimidine**
	1,4-Diazine **Pyrazine**		1,4-Perhydrodiazine **Piperazine**
	4*H*-Oxin **4*H*-Pyran**		2*H*-Oxin **2*H*-Pyran**
	Oxane **Tetrahydropyran**		1,4-Perhydroxazine **Morpholine**

Fig. 2.35: Selected six-membered heterocyclic skeletons.

requires localizers for hetero atoms (e.g. 1,3,4-triazole), and these numbers do not correspond to those of the system after fusion, they should be indicated in square brackets, apart from the localizers for fusion, which also go in a pair of square brackets.

When the secondary component to be named is furan, thiophene, pyridine, quinoline, isoquinoline, pyrimidine, or imidazole, the terms "furo", "thieno", "pyrido", "quino", "isoquino", "pyrimidino", and "imidazo", respectively, are used.

Finally, if a heteroatom is at the confluence of two rings, separation of the system into components will be considered part of the two that result (Fig. 2.36).

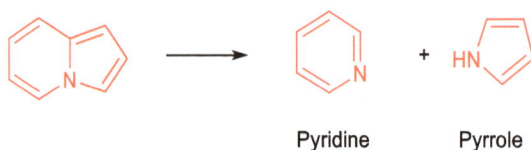

Pyridine   Pyrrole

Fig. 2.36: Separation of a bicyclic system into its components, when the heteroatom is at the confluence of two rings.

2.11 Numbering of condensed heterocycles

The numbering of the condensed heterocycle follows the same orientation and start operations in the upper right ring as described for nonheterocyclic systems. The heteroatoms common to two or more rings now have their own locator, a number that is not followed by a letter, as in the carbocyclic systems. When more than one correct orientation is possible, the following criteria are applied until the ambiguity is eliminated:
(a) Lowest numbering to the heteroatoms together
(b) Lowest numbering for the most preferred heteroatom
(c) Lowest numbering for carbons common to rings
(d) Lowest numbering to hydrogenated positions

First, the common link is located, and then the simple heterocycles that compose it are named (simple component is one that has a common name but with the most complex structure possible).

The heterocycle, which gives the compound the name, is the heterocycle base, which is preceded by a bracket with two numbers and a letter indicating the common link, and before the bracket the non-base heterocycle is placed as a prefix. The lowercase letter locates the common link of the base heterocycle, and the numbers locate the non-base heterocycle common bond. The sense traversed by the base heterocycle is then matched to the path of the non-base heterocycle. Finally, the complete condensed heterocycle is numbered.

The most commonly used prefixes are furan, imidazo (imidazole), isoquino (isoqui-noline), pyrido (pyridine), pyrazine (pyrazine), pyran (pyran), pyrimido (pyrimidine), quinoline (quinoline), thieno (thiophene), pyrrolo (pyrrole), and pyrazole (pyrazole).

We will number the following condensed heterocycle as an example (Fig. 2.37).

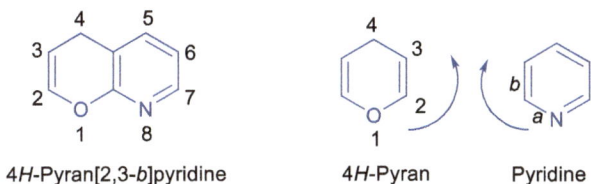

4H-Pyran[2,3-b]pyridine   4H-Pyran   Pyridine

Fig. 2.37: Naming and numbering a condensed bis-heterocycle, having a nitrogen atom.

2.12 Criteria for choosing the base heterocycle

1.   The one containing nitrogen (pyridine versus pyran)
2.   If there is no nitrogen, the one containing the heteroatom other than N, in the order of preference set ($O > S$) (Fig. 2.38)

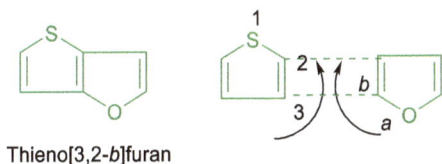

Thieno[3,2-b]furan

Fig. 2.38: Naming and numbering a condensed bis-heterocycle, not having a nitrogen atom.

3.   The one with the highest number of cycles in the system (Fig. 2.39)

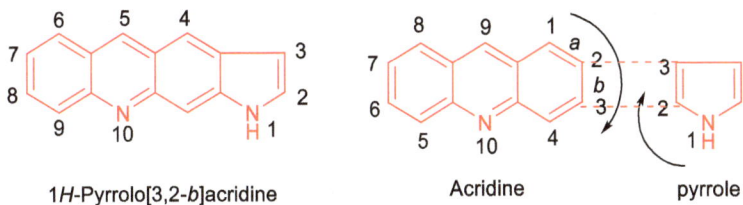

1H-Pyrrolo[3,2-b]acridine   Acridine   pyrrole

Fig. 2.39: The highest number of cycles is the main component for a condensed heterocycle.

4.   The one with the largest ring (Fig. 2.40)

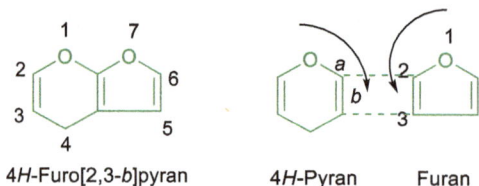

4H-Furo[2,3-b]pyran   4H-Pyran   Furan

Fig. 2.40: The largest ring is the main component for a condensed bis-heterocycle with the same heteroatom.

5. A component containing a heteroatom with the higher priority ($O > S > N$) (Fig. 2.41)

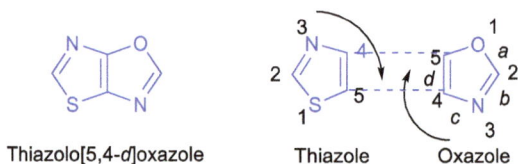

Thiazolo[5,4-d]oxazole   Thiazole   Oxazole

Fig. 2.41: The component containing a heteroatom with the higher priority ($O > S > N$) is the main component for a condensed bis-heterocycle with the same size.

6. Of the two components that have the same size, number, and type of heteroa-tomic bases, the one that is considered to have the lower numbering for the het-eroatoms is chosen as the prefix (Fig. 2.42)

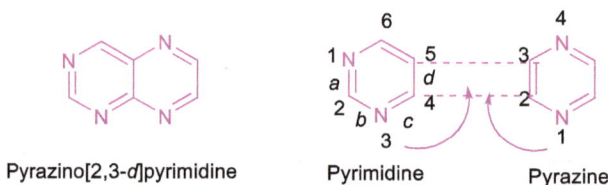

Pyrazino[2,3-d]pyrimidine   Pyrimidine   Pyrazine

Fig. 2.42: The component with the lower numbers for the heteroatoms before fusion is the main component for a condensed bis-heterocycle with the same heteroatom.

Consider the following examples. In the case of doxepin, the central heterocycle is an oxepin, partially hydrogenated and condensed with two benzenes (Fig. 2.43).

Thus, it is a dibenzo[b,e]oxepin. Their numbering must match the lowest possible number (5) to the oxygen. Its systematic name is 11-[3-(dimethylamino)propylidene]-6,11-dihydrodibenzo[b,e]oxepin or alternatively as (E)-3-(dibenzo)[b,e]oxepin-11(6H)-ylidene)-N,N-dimethylpropan-1-amine, taking into account the amino function is the main functional group (Fig. 2.43).

Clothiapine raises another problem (Fig. 2.44).

Doxepin
(antidepressant agent)

Numbering and orientacion

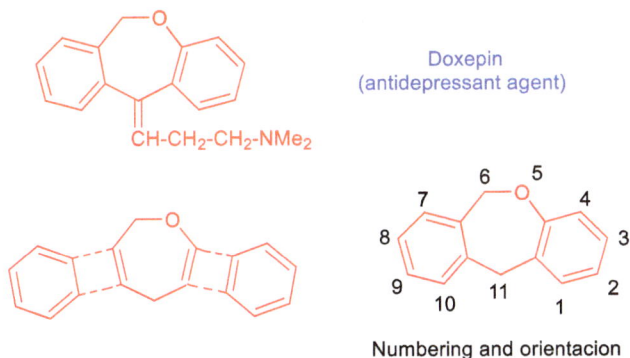

Fig. 2.43: Numbering and orientation of doxepin.

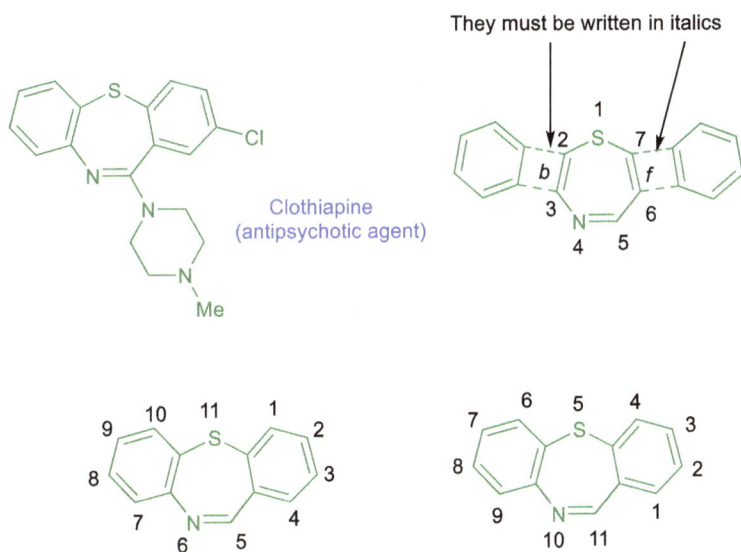

They must be written in italics

Clothiapine
(antipsychotic agent)

Fig. 2.44: Numbering and orientation of clothiapine.

The second numbering is correct, since sulfur is preferential in relation to nitrogen. Since the positions of S and N after condensation (5 and 10) do not coincide with their locators in 1,4-thiazepine, these should be indicated in brackets: 2-chloro-11-(4-methyl -1-piperazine-1-yl)dibenzo[b,f]-1,4-thiazepine.

2.13 Heterocycles condensed with benzene

They are named by placing the "benzo" prefix in the name of the heterocycle, losing the final "o" if the heterocycle begins with a vowel. The position of the heteroatoms in the condensed end system is indicated by locators, and with a letter the side by which it binds to benzene. Examples are shown in Fig. 2.45.

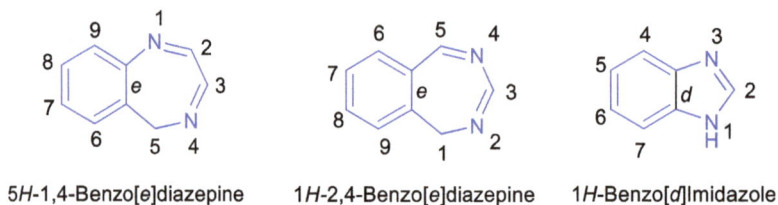

5*H*-1,4-Benzo[*e*]diazepine   1*H*-2,4-Benzo[*e*]diazepine   1*H*-Benzo[*d*]Imidazole

Fig. 2.45: Heterocycles condensed with benzene.

After Fig. 2.46, where some of the most important heterocyclic systems are selected, a couple of examples of drugs will be proposed, in which several aspects treated in this chapter are consolidated.

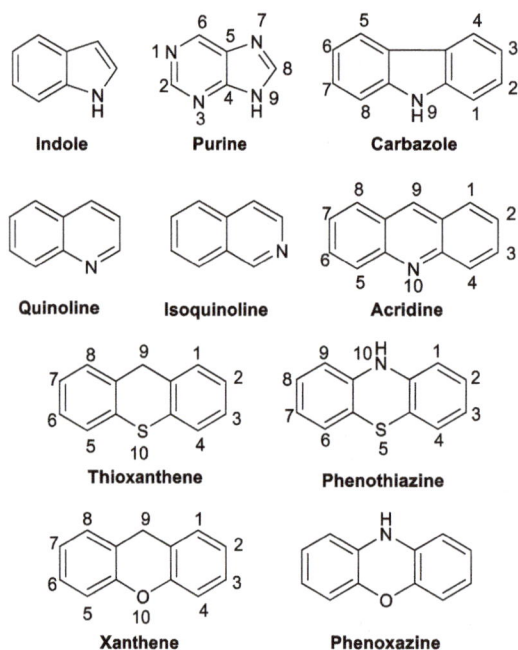

Indole   Purine   Carbazole

Quinoline   Isoquinoline   Acridine

Thioxanthene   Phenothiazine

Xanthene   Phenoxazine

Fig. 2.46: Names of some condensed heterocyclic systems selected, in many cases, with a specific numbering system.

Consider the case of a phenothiazine with a substituent at its 10-position, which contains a spiro derivative (Fig. 2.47).

Finally, let us look at another example in the area of benzodiazepines. The first is the hypnotic-sedative drug estazolam (Fig. 2.48).

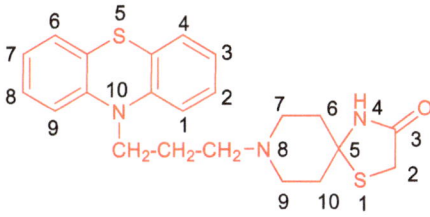

8-[3-(Phenothiazin-10-yl)propyl]-1-thia-4,8-diazaspiro[4,5]decan-3-one

Fig. 2.47: A phenothiazine with a substituent that contains a spiro derivative.

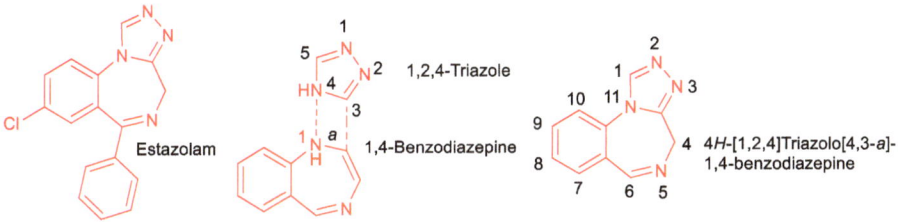

Fig. 2.48: Name and numbering of estazolam.

2.14 Exercises

A. Name the following drugs:

(1)

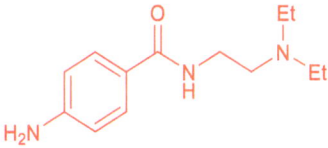

(2)

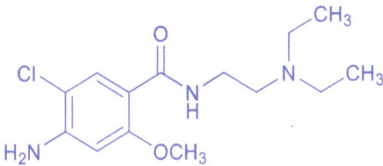

(3)

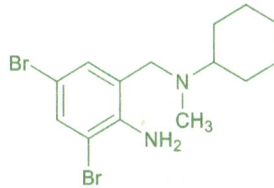

(4)

(5)

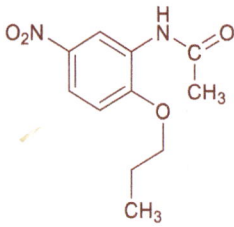

(6)

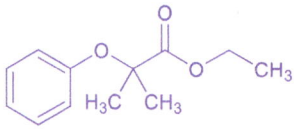

(7)

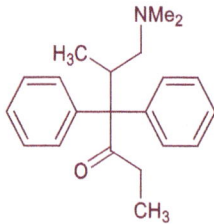

(8)

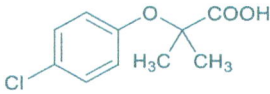

(9)

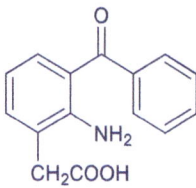

(10)

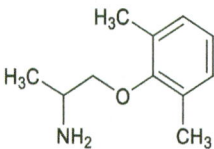

(11)

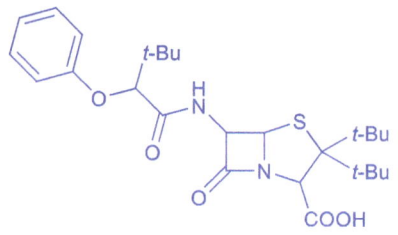

(12)

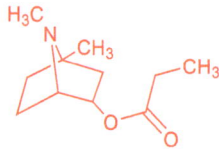

(13)

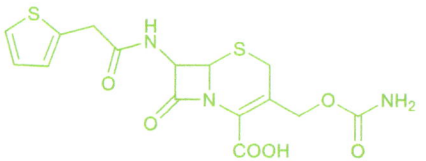

(14)

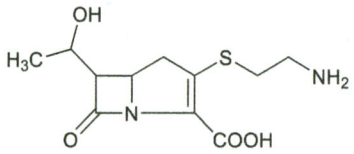

(15)

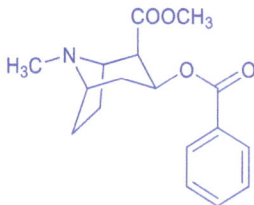

(16)

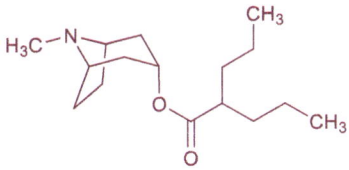

(17)

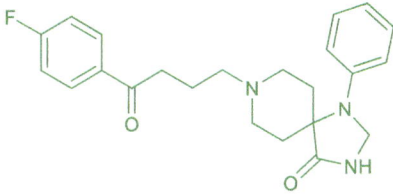

(18)

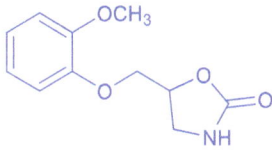

(19)

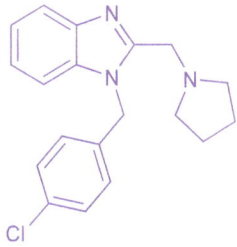

(20)

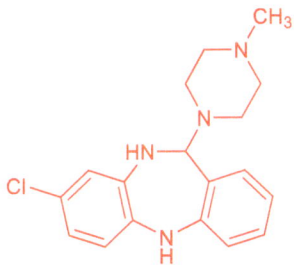

(21)

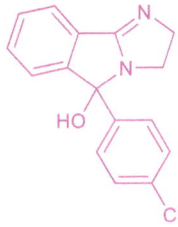

(22)

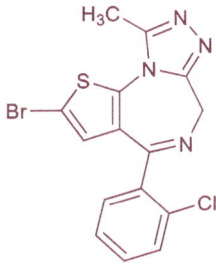

(23)

(24)

(25)

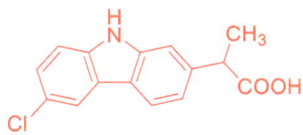

(26)

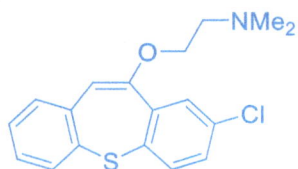

(27)

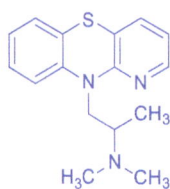

(28)

(29)

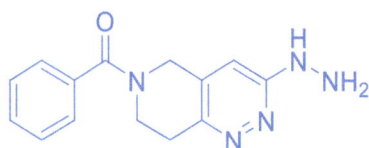

(30)

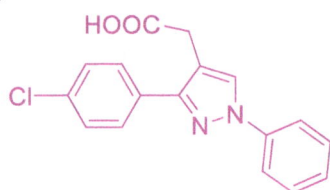

(31)

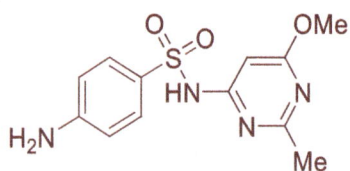

(32)

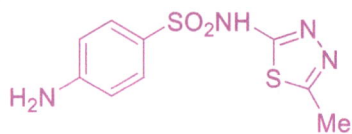

(33)

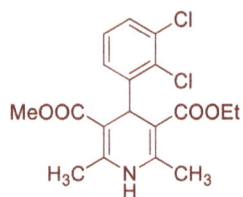

(34)

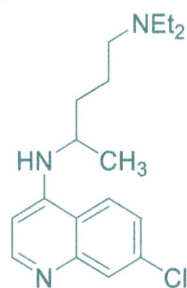

(35)

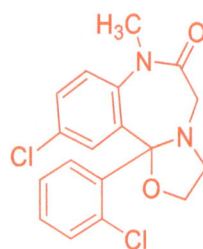

(36)

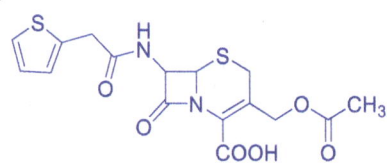

B. Formulate the following drugs:
 (1) 2-(4-Phenyl-3-fluorophenyl)propionic acid
 (2) 2-[2-(2,6-Dichlorophenylamino)phenyl]acetic acid
 (3) Ethyl 7-chloro-4-ethoxy-6-fluoroquinoline-3-carboxylate
 (4) 7-Chloro-10-(2-dimethylaminoethyl)-5,10-dihydrodibenzo[*b,e*]-1,4-diazepine-11-one
 (5) 3-(4,5-Diphenyloxazol-2-yl)propionic acid
 (6) 5-(3-Methylamino)propyl-5*H*-dibenzo[*b,f*]azepine; alternatively named as 3-(5*H*-dibenzo[*b,f*]azepine-5-yl)-*N*-methylpropan-1-amine
 (7) 10-(3-Dimethylaminopropyl)-9,9-dimethyl-9,10-dihydroacridine; alternatively named as 3-(9,9-dimethylacridin-10(9*H*)-yl)-*N,N*-dimethylpropan-1-amine
 (8) Ethyl-7-chloro-2,3-dihydro-2-oxo-5-phenyl-1*H*-benzo[*e*]-1,4-diazepine-3-carboxylate
 (9) 5-(Dimethylamino)-9-methyl-2-propyl-1*H*-benzo[*e*]pyrazolo[1,2-*a*]-1,2,4-triazine-1,3(2*H*)-dione
 (10) 3-Acetoxy-5-(2-diethylaminoethyl)-2-(4-methoxyphenyl)-2,3-dihydro-1,5-benzothiazepine-4(5*H*)-one; alternatively named as 8-chloro-5-(2-(diethylamino)ethyl)-2-(4-methoxyphenyl)-4-oxo-2,3,4,5-tetrahydrobenzo[*b*]-1,4-thiazepine-3-yl acetate
 (11) 5-(2-Fluorophenyl)-1-methyl-7-nitro-1,3-dihydrobenzo[*e*]-1,4-diazepine-2-one
 (12) 2-Methyl-3-(2-methylphenyl)quinazoline-4(3*H*)-one
 (13) 7-[(2-Amino-2-phenyl)acetamido]-3-methyl-8-oxo-5-thia-1-azabicyclo [4.2.0]oct-2-ene-2-carboxylic acid
 (14) 11-Chloro-12*b*-phenyl-2,8-dimethyl-8,12b-dihydro-5*H*-[1,3]oxacino[3,2-*d*]-1,4-benzodiazepine-4,7-dione; alternatively named as 11-chloro-2,8-dimethyl-12*b*-phenyl-8,12*b*-dihydro-4*H*-benzo[*f*]-1,3-oxazino[3,2-*d*]-1,4-diazepine-4,7(6*H*)-dione
 (15) 6-(2-Phenoxybutanamido)-3,3-dimethyl-7-oxo-4-thia-1-azabicyclo[3.2.0]heptane-2-carboxylic acid
 (16) 2-Butyryl-10-{3-[4-(2-hydroxyethyl)-1-piperazinyl]propyl}phenothiazine; alternatively named as 1-{10-[3-(4-(2-hydroxyethyl)piperazine-1-yl]propyl}-10*H*-phenothiazine-2-yl)butan-1-one
 (17) 4-[4-(4-Chlorophenyl)-4-hydroxypiperidine-1-yl]-1-(4-fluorophenyl)butan-1-one
 (18) *N*-(4-Chlorobenzyl)-*N',N'*-dimethyl-*N*-pyridine-2-yl-ethylene-1,2-diamine; alternatively named as N^1-(4-chlorobenzyl)-N^2,N^2-dimethyl-N^1-(pyridine-2-yl)ethane-1,2-diamine
 (19) 6-(2,6-Dimethoxybenzamido)-3,3-dimethyl-7-oxo-4-thia-1-azabicyclo[3.2.0]heptane-2-carboxylic acid
 (20) 1-[4-Hydroxy-3-(hydroxymethyl)phenyl]-2-*tert*-butylaminoethanol

3 Search for prototypes

3.1 Goals

- To know the global process through which new prototypes are discovered and the methodologies used over time
- To know the importance of isolation, purification, and structural determination of drugs

3.2 Traditional and current discovery of new drugs

One of the greatest difficulties that the student finds when studying pharmaceutical chemistry (PC) is both the high number and the great structural diversity of the drugs to be memorized. Therefore, in order to facilitate the learning of this discipline, the authors have decided to minimize the number of structures and, in addition, the chemical structures detailed in these first general chapters will be later studied in the corresponding descriptive themes in Volume 2 of this series.

The first thing required to begin a PC project is to have a leader, that is, a compound that can be considered as therapeutically useful. The level of biological activity may not be particularly noticeable, because this is not a fundamental issue. The lead compound is not intended for use in clinical practice. It is only the starting derivative from which a clinically useful compound can be developed. It does not matter whether or not this compound is toxic or has side effects. One of the fundamental objectives of PC is the search for drugs that are more potent, more selective, and less toxic in their therapeutic action.

Lead compounds can be obtained from a variety of natural sources, such as flora and fauna, or synthetic compounds prepared in the laboratory. If a drug or poison produces a biological effect, it is because there must be a molecular target in the body. Many of the earliest known drugs, such as the analgesic morphine (Fig. 3.1), are natural compounds derived from plants that coincidentally interacted with a molecular target of the human body. Subsequently, chemical messengers or neurotransmitters began to be discovered. These have a short life span and are released by the nerves to interact with specific cellular targets. For example, since the 1970s, a wide variety of peptides and proteins that act as natural analgesics (enkephalins and endorphins) of the body have been discovered. Quinine is a natural alkaloid with antipyretic, antimalarial, and analgesic properties.

However, few chemical messengers have been identified, either because they are present in very small amounts or because they have such a short half-life that their isolation is not possible. In fact, today many chemical messengers are unknown. This implies that many of the potential biological targets of the human body remain hidden. However, advances in genomics and proteomics have changed the whole picture

https://doi.org/10.1515/9783111316901-003

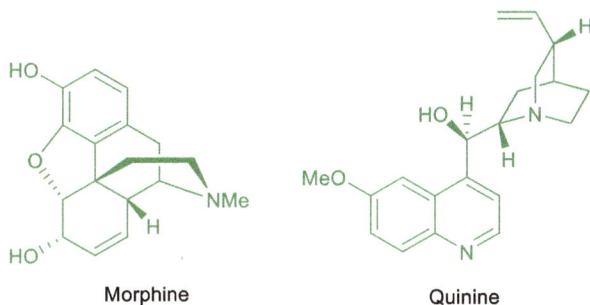

Fig. 3.1: Lead compounds such as morphine and quinine.

of new drug development. Several genome projects have elucidated the DNA of humans and other life forms and have discovered a large number of new proteins that are biological targets for the future.

These biological targets have been hidden for so long that their natural chemical messengers are also unknown, and for the first time, PC is facing new targets without leading compounds that interact with them. Such targets have been defined as orphan receptors, and the current challenge is to find chemical compounds that interact with each of these targets to discover what their functions are and to see if they are appropriate as drug targets.

3.2.1 Screening of natural products

With this search procedure, a large number of chemical compounds, natural or synthetic, are subjected to a battery of pharmacological tests, in search of an unknown or hypothetical action. This mass trial is one of the most costly and least active procedures; however, it has been successfully used in some fields, such as in the discovery of new antibiotics like chloramphenicol (Fig. 3.2), isolated from *Streptomyces venezuelae* in the general microorganism screening program at Parke-Davis Laboratories in the USA.

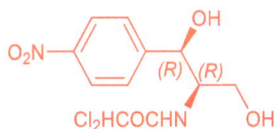

Fig. 3.2: Chloramphenicol.

Natural products are an important source of biologically active compounds. Generally, the natural source has some form of biological activity, and the compound responsible for this activity is known as the active principle. This structure can act as a leader compound. Most biologically active natural products are secondary metabolites with complex structures. Unfortunately, this complexity makes its synthesis difficult, and the compound generally has to be extracted from its natural source, a slow, expensive, and inefficient process. Therefore, the design of simpler compounds is usually an effective approach for the development of new drugs.

Plants have always been a rich source of leading compounds (e.g. morphine, cocaine, quinine, tubocurarine, nicotine, and muscarin). Many of them are useful drugs in themselves (e.g. morphine and quinine), and others have been the basis for synthetic drugs (e.g. local anesthetics developed from cocaine).

Microorganisms such as bacteria and fungi have also been the source of drugs and leading compounds. The screening of microorganisms became very popular after the discovery of penicillin. Land and water samples have been collected from all over

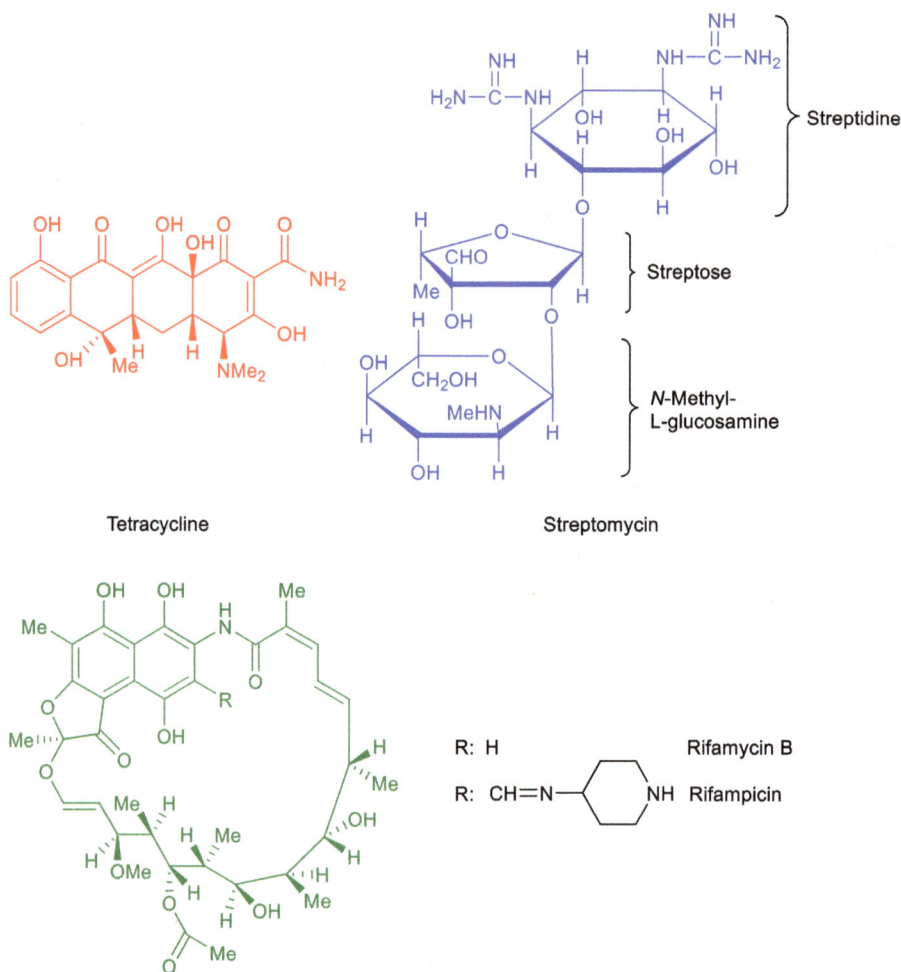

Fig. 3.3: Naturally existing or semisynthetic antibiotics.

the world to study new strains of fungi and bacteria and have resulted in a huge arsenal of antibacterial agents such as cephalosporins, tetracyclines, aminoglycosides, rifamycins (Fig. 3.3), and chloramphenicol.

Poisons and toxins have been used as lead compounds in the development of new drugs. For example, the teprotide (nonapeptide isolated from the Brazilian viper *Bothrops jararaca*) was the leading compound for the development of the antihypertensive agent captopril. Teprotide contains the strange pyroglutamic acid in its N-terminal part and the proline in its C-terminal part (Fig. 3.4):

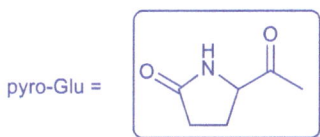

Pyro-Glu-Trp-Pro-Arg-Pro-Glu-Ile-Pro-Pro-OH

pyro-Glu =

Pyroglutamic acid is an uncommon natural amino acid derivative
in which the free amino group of glutamic acid or glutamine cyclizes to form a lactam

Fig. 3.4: Teprotide.

3.2.2 Existing drugs used as leaders

The molecular manipulation of known drugs consists of a progressive modification of the chemical structure of chemical substances that possess a certain biological activity. Many companies use drugs established by their competitors as leading compounds to design a drug that allows them to establish themselves in the same market area. The aim is to modify the structure sufficiently to avoid, on the one hand, the restrictions of the patent, and on the other hand to improve the therapeutic properties. For example, the antihypertensive captopril was used as a lead compound to produce their own antihypertensive agents (Fig. 3.5).

Although often neglected as "twin" drugs, they sometimes show improvements over the original drug (better drugs: "me better"). For example, modern penicillins are more selective, more potent, and more stable than the original penicillins.

An existing drug may have another minor property or an undesirable effect that may be useful in another area of medicine. In this way, the drug could act as a leader in line of its side effects. Consequently, the aim would be to increase the side effect and eliminate the major biological activity. This has been described as the "selective optimization of side activities" approach.

For example, most sulfonamides have been used as antibacterial agents. However, some sulfonamides cannot be used clinically because they produce hypoglycemia and accordingly cause seizures. Accordingly, structural modifications were performed to eliminate the antibacterial activity and potentiate hypoglycemic activity. This led to the

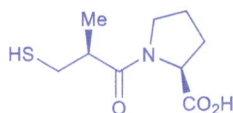

Captopril

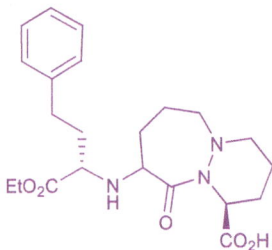

Cilazapril
(Hoffman-LaRoche)

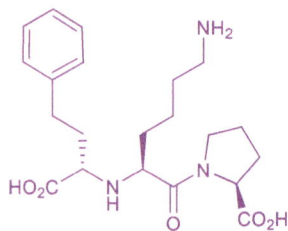

Lisinopril
(Merck)

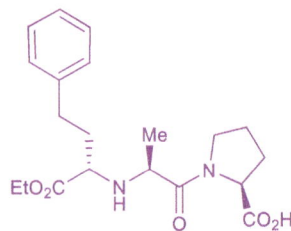

Enalapril
(Merck)

Fig. 3.5: Captopril and "me too" drugs.

antidiabetic agent tolbutamide (Fig. 3.6). Similarly, the development of sulfonamide-type diuretics such as chlorothiazide arose from the observation that sulfanilamide has a diuretic effect at high doses (due to its action on the carbonic anhydrase enzyme).

In some cases, the side effect may be strong enough for the drug to be used without modification. For example, the drug sildenafil (Viagra®) that treats male impotence (Fig. 3.6) was originally designed as a vasodilator to treat angina and hypertension.

Chlorpromazine is used as a neuroleptic (a neuroleptic or antipsychotic is a drug that is commonly, but not exclusively, used for the treatment of psychosis) in psychiatry, but developed from the antihistamine agent promethazine. It may seem strange, but it was known that promethazine has sedative effects, so pharmaceutical chemists modified the structure to increase the sedative effects at the expense of antihistamine activity.

3.2.3 Isolation and identification of drug metabolites

Acetalinide and acetophenidine are converted in the body in paracetamol (International Nonproprietary Name), which is actually the active drug (analgesic–antipyretic) in cases of intolerance to Aspirin® (Scheme 3.1).

Sulfanilamide

Sulfonamide analogs
R is usually an aromatic
heterocyclic derivative

Tolbutamide

Chlorothiazide

Sildenafil (Viagra®)

Promethazine

Chlorpromazine

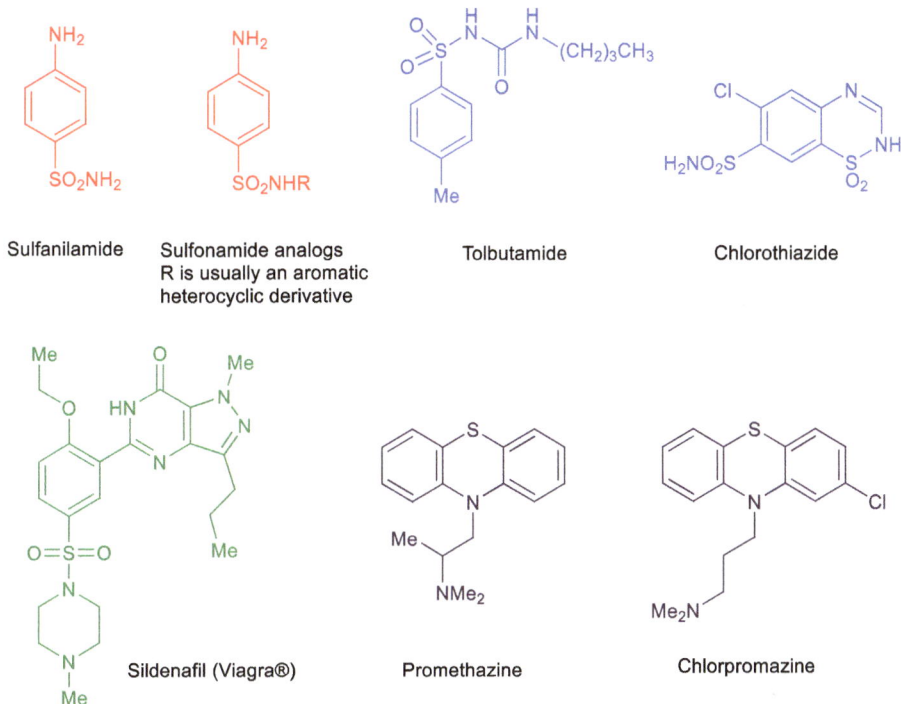

Fig. 3.6: Drugs developed through potentiation of a side effect.

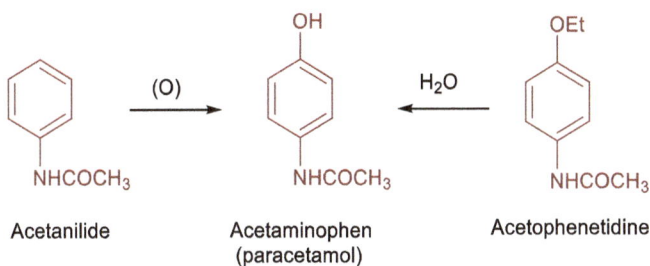

Acetanilide

Acetaminophen
(paracetamol)

Acetophenetidine

Scheme 3.1: Relationship between paracetamol and its two metabolic precursors.

3.2.4 Examples of serendipity

The term "serendipity" is used more and more in science to describe the way in
which many of the great discoveries have been made. Chance plays a very important
role in unforeseen valuable findings, when luck favors the prepared mind, in the
words of Pasteur (one of the beneficiaries of serendipity). This term can be defined as
"the ability to make discoveries by accident and sagacity, when one is looking for

something else". It was 1754 when the English writer Horace Walpole encountered an ancient story from Asia, which occurred in Serendip, a former name of the present country of Sri Lanka, also previously known as the Kingdom of Ceylon. The monarch of that country sent his three children to travel the world to get the essential items he needed. The princes of Serendip visited the villages, had remarkable experiences, and returned with far more valuable finds than their father had asked them. Serendipity was decisive in the discovery of America, since we know that what Columbus was looking for was a shorter way to reach the Indies.

Some notable examples are depicted in Scheme 3.2, and in Figs. 3.7 and 3.8.

(a) For example, *prontosil rubrum*, which is initially synthesized as an azo dye, is useful for staining microorganisms, casually opening the field of chemotherapy when tested in vivo and demonstrated its antibacterial therapeutic activity (Scheme 3.2).

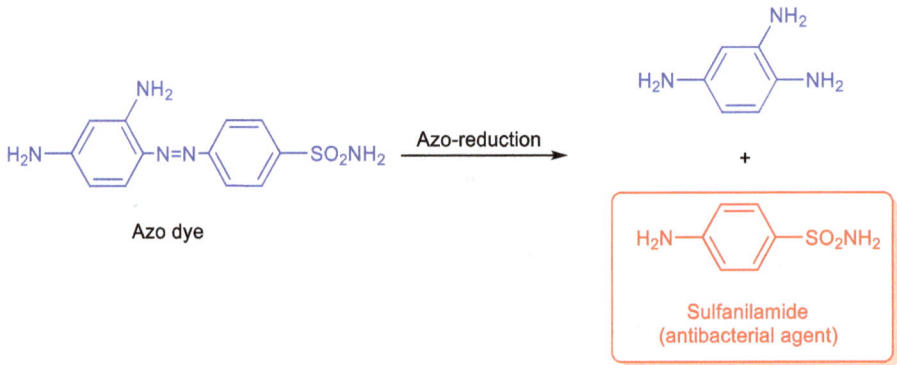

Scheme 3.2: Examples of metabolic reduction.

(b) Rubber industry workers were repelled by alcohol. This was caused by an antioxidant agent, which was used in the rubber manufacturing process. Ethanol is metabolized by sequential metabolic oxidation, first to acetaldehyde by alcohol dehydrogenase and then to acetic acid by means of aldehyde dehydrogenase. Disulfiram irreversibly inhibits the oxidation of acetaldehyde, causing a considerable increase in acetaldehyde concentrations after the ingestion of alcohol, which was so unpleasant that workers preferred not to drink. The antioxidant agent became the leading compound for the development of disulfiram (Antabuse®), used for the treatment of chronic alcoholism. It is believed that the ability of disulfiram to react with the sulfhydryl groups of essential proteins (generating diethyldithiocarbamate) is important for its activity (Fig. 3.7).

(c) Clonidine was initially designed as a nasal vasoconstrictor. Clinical trials revealed that this this effect is due to a marked decrease in blood pressure, which makes it an important antihypertensive agent.

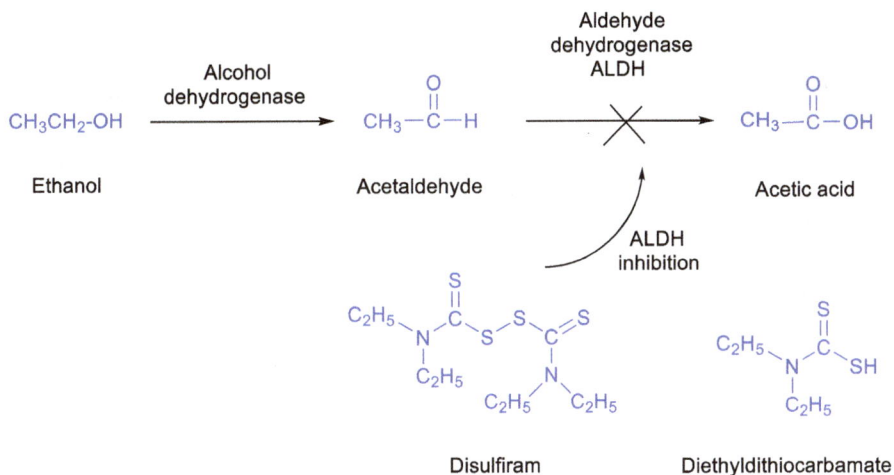

Fig. 3.7: Inhibition of aldehyde dehydrogenase (ALDH) by disulfiram.

(d) Imipramine was synthesized as an analog of chlorpromazine and was initially used as an antipsychotic agent. It was found, however, to alleviate depression, which led to the development of a number of compounds classified as tricyclic antidepressants (Fig. 3.8).

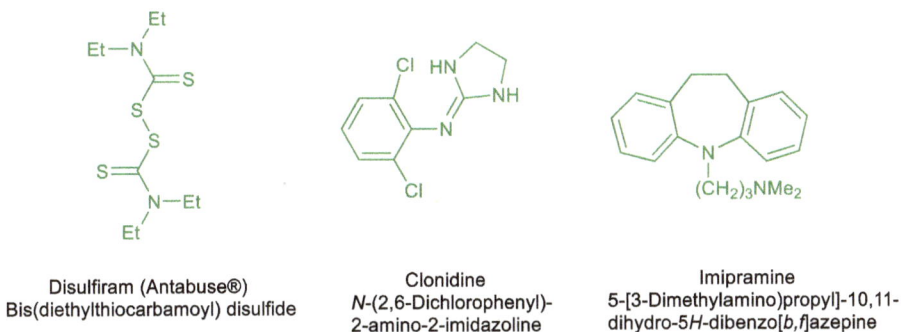

| Disulfiram (Antabuse®) Bis(diethylthiocarbamoyl) disulfide | Clonidine *N*-(2,6-Dichlorophenyl)- 2-amino-2-imidazoline | Imipramine 5-[3-Dimethylamino)propyl]-10,11- dihydro-5*H*-dibenzo[*b,f*]azepine |

Fig. 3.8: Examples of drugs discovered by serendipity.

3.3 Planned syntheses of new chemical compounds on rational bases

It is the golden dream of pharmaceutical chemists and pharmacologists. The ultimate aim in the synthesis of a drug is to prepare one for the "measurement" of the biological function to be modified.

3.4 Isolation and purification

If the compound (or active principle) is present in a mixture of other compounds, it must then be isolated and purified. The ease with which the active ingredient can be purified and isolated depends largely on its structure, stability, and the amount of the compound. For example, Fleming recognized the antibiotic properties of penicillin and its nontoxic characteristics for humans, but did not consider it for clinical use, because he was not able to purify it. He could isolate it in an aqueous solution, but whenever he tried to eliminate the water, the drug was decomposed. It was not until the development of new experimental procedures such as lyophilization and chromatography that the isolation and purification of both penicillin and other natural products became possible.

3.5 Structural determination

Nowadays, a structural determination can take a week of work but in the past, it could be carried out for two to three decades. For example, the empirical formula of cholesterol was known in 1888, but its chemical structure was not completely established until 1932 by X-ray crystallography (Fig. 3.9).

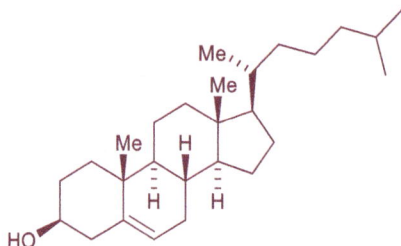

Fig. 3.9: Cholesterol.

In the past, structures were degraded to simpler compounds, which could be further degraded to readily recognizable fragments. From these smaller fragments, a possible structure was proposed, but the only way to test the proposed structure was to carry out its total synthesis and compare its physical and chemical properties with those of the natural compound.

Today, structural determination is a relatively simple process, and it is only necessary to carry out its total synthesis to corroborate its structure when it has been obtained in minimal quantities. The most useful analytical techniques for structural determination are X-ray crystallography and NMR spectroscopy. The application of the first technique allows obtaining a "snapshot" of the molecule, but a suitable crystal of the molecule is required. The latter technique is used more routinely and can be carried out on any sample, whether solid, oily, or liquid. There are different NMR experiments that can be used to establish the structure of complex molecules, such as two-dimensional NMR techniques.

In cases where there is not enough sample for an NMR analysis, mass spectrometry can be very useful. The fragmentation model can provide important clues about the structure, but it does not definitively prove the chemical structure. In this case, complete synthesis would be required as the final test.

3.6 Key notes

1. A prototype is a structure that shows a useful pharmacological activity and can act as the starting point for drug design.
2. Natural products are a rich source of hit compounds. The agent responsible for the biological activity of a natural extract is known as the active principle.
3. Serendipity has played a role in the discovery of new compounds.
4. If a hit compound is present in a natural extract, it has to be isolated and purified in order to determine its structure. X-ray crystallography and NMR spectroscopy are particularly important in its structure determination.
5. A molecular target is chosen, based on the fact that is believed to influence a particular disease, when affected by a drug. The greater the selectivity that can be achieved, the less chance of side effects.

3.7 A pitfall

It is possible to identify whether a particular enzyme or receptor plays a role in a particular ailment. However, the body is a highly complex system. For any given function, there are usually several messengers, receptors, and enzymes involved in the process. For example, there is no one simple cause for hypertension. This is illustrated by the variety of receptors and enzymes, which can be targeted in its treatment. These include β_1-adrenoreceptors, calcium ion channels, angiotensin-converting enzyme, and the potassium ion channel. As a result, more than one target may need to be addressed for a particular ailment. For example, most of the current therapies for asthma involve a combination of a bronchodilator (β_2-agonist) and an anti-inflammatory agent such as a corticosteroid. In the medical practice, a cocktail of drugs is typically used.

4 Optimization of prototypes

4.1 Goals

- To know the different tools of structural variation and their use in prototype optimization in the discovery of new drugs
- To start with the concept of the cost of research: the idea that starting from active compounds it is possible to produce other active compounds is an economic concept

4.2 Molecular modification

Molecular manipulation consists of a progressive modification of the chemical structure of substances that possess a certain biological activity (prototypes or series heads); it is the most frequent and, so far, the most cost-effective procedure for the genesis of new drugs. About 80–90% of the current drugs have been obtained in this way.

Since there is a high probability that a molecule obtained by modification of an active prototype has useful properties, this drug-screening process is usually more productive than the assay, without a sufficiently solid base, of novel compounds isolated from nature or synthesized at the laboratory. In addition, it offers economic advantages, since both the synthetic methods and the pharmacological tests of the analogs will be similar to those used for the reference compound.

4.2.1 Purpose

- Development of substitutes and therapeutic copies
- Development of drugs with a different action spectrum
    - Transformation of agonists into antagonists
    - Separation of one of the components of the action
- Modification of pharmacokinetics
- Modification of the distribution
- To increase chemical stability

4.2.1.1 Development of substitutes and therapeutic copies
(1) Development of existing drug substitutes, more powerful if possible. Examples:
- The passage of pronetalol (β-adrenergic blocker) to propranolol, with greater potency than the previous one (Fig. 4.1).

https://doi.org/10.1515/9783111316901-004

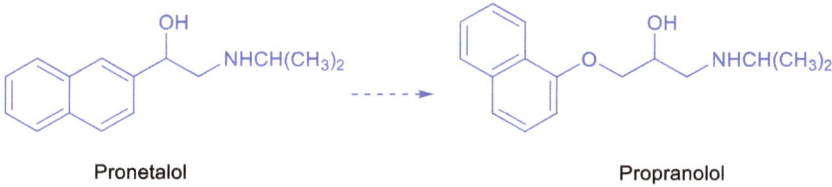

Fig. 4.1: Transition from pronetalol to propranolol.

– Conversion of (*S*)-1-(3-mercaptopropionyl)-L-proline into captopril {(2*S*)-*N*-[3-mercapto-2-methylpropionyl]-L-proline}, an antihypertensive agent, ten times more potent than its demethylated analog (Fig. 4.2).

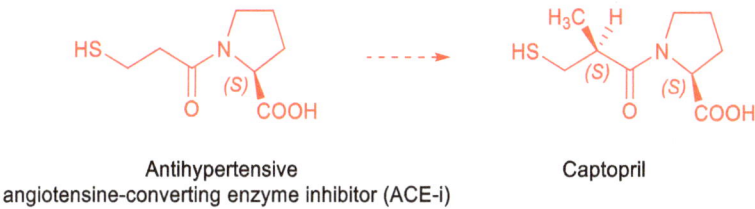

Antihypertensive
angiotensine-converting enzyme inhibitor (ACE-i)

Captopril

Fig. 4.2: Transition from (*S*)-1-(3-mercaptopropionyl)-L-proline into captopril.

(2) The development of therapeutic copies is a quick option to find new patentable formulas that can successfully compete in the market. They require less economic investment but must compete with existing drugs of proven efficacy in the same therapeutic group. Their commercial success is conditioned by the galenic or therapeutic advantages that they provide with respect to existing drugs. From the antihypertensive drug enalapril, compounds such as lisinopril and cilazapril were developed (Fig. 4.3).

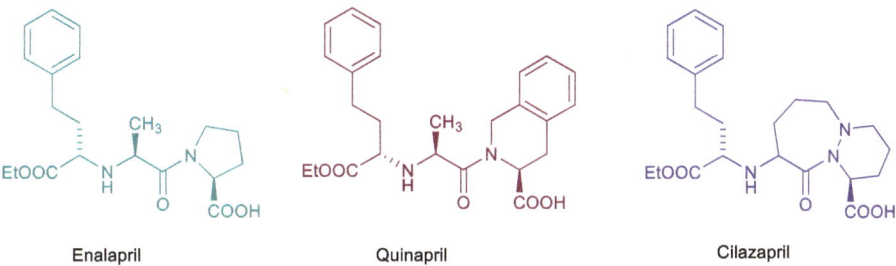

Enalapril

Quinapril

Cilazapril

Fig. 4.3: Development of therapeutic copies.

4.2.1.2 Development of drugs with different spectrum of action

(1) Transformation of agonists into antagonists. Thus, the β-adrenergic antagonist dichloroisoprenaline is obtained from the isoprenaline agonist by substituting the phenolic hydroxyl groups for chlorine atoms (Fig. 4.4).

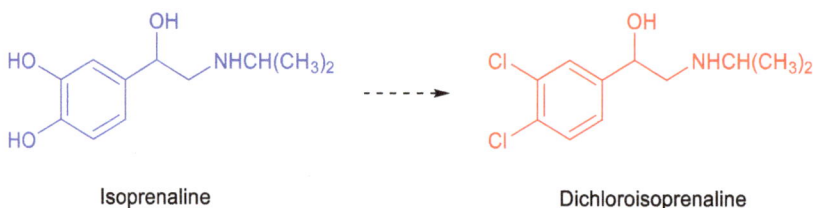

Isoprenaline   Dichloroisoprenaline

Fig. 4.4: Development of an antagonist (dichloroisoprenaline) from an agonist (isoprenaline).

(2) Separation of any component to enhance or modify the action of the prototype. Examples:

- The virilizing effect of testosterone is lost when it is transformed into a 19-noresteroid, norethandolone, which is used as an anabolic drug (Fig. 4.5).

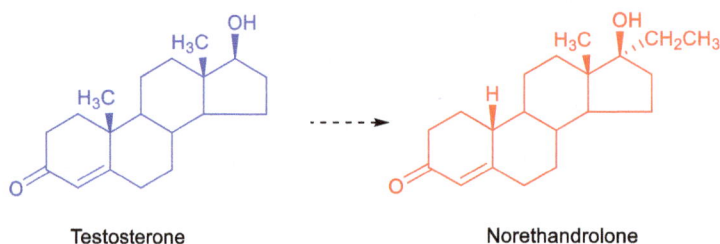

Testosterone   Norethandrolone

Fig. 4.5: Norethandrolone that has not the virilizing action of testosterone.

- The weak uricosuric character of phenylbutazone, attributed to the acidity of the hydrogen at 4 position, can be enhanced by substituting its butyl group for an electron-withdrawing phenylsulfinylethyl group, which reinforces the acidic character, resulting in sulfinpyrazone (more potent uricosuric) (Fig. 4.6).

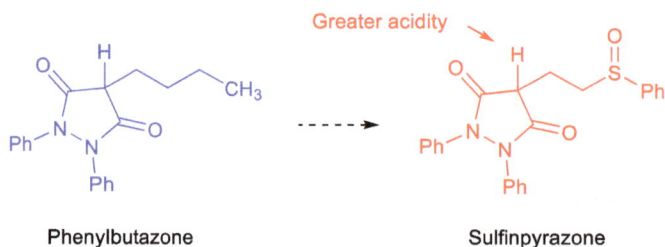

Phenylbutazone   Sulfinpyrazone

Fig. 4.6: Sulfinpyrazone has a higher uricosuric activity than phenylbutazone.

4.2.1.3 Modification of pharmacokinetics

The dose–effect relationship can be modified with structural changes that alter the rate of metabolism and/or excretion. Example: passage of tolbutamide, oral antidiabetic of short duration of action, to clorpropamide of a more lasting action (Fig. 4.7).

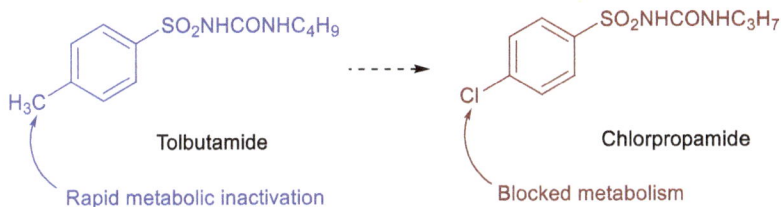

Fig. 4.7: From an oral short-acting antidiabetic to another one with a more acting effect.

4.2.1.4 Modification of distribution

This allows the drug to reach the different organs and tissues in which it acts. This purpose can be achieved through various procedures:

1. Introducing a fixed ionic charge to prevent the passage through the blood–brain barrier (Fig. 4.8).

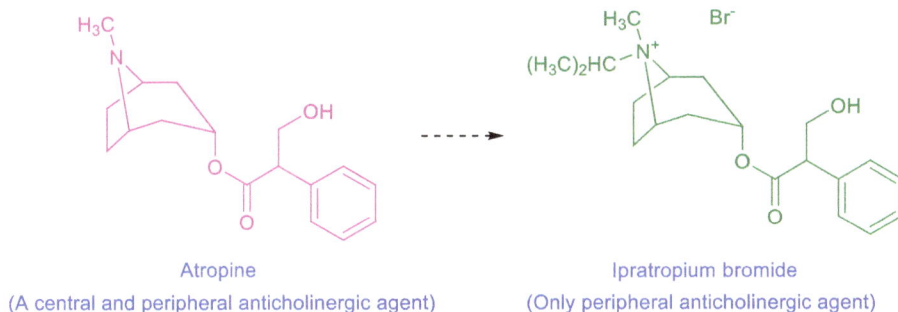

Fig. 4.8: Introducing a fixed ionic charge to prevent the passage through the blood–brain barrier (BBB).

2. Using carrier groups to increase cell or tissue selectivity of the drug.
The active structure of nitrogen mustards, nonselective alkylating agents, can be transformed into useful anticancer agents by increasing their selectivity by rapidly dividing cells. To this end, fixed conveyor groups are introduced on the N, which facilitate the active transport of the thus latent drug to the sites of action. Transporting groups such as L-phenylalanine or uracil, present in melphalan and uramustine, can be used, because tumor tissues synthesize proteins and DNA at a faster rate than healthy tissues so that their demand for amino acids and bases is greater. Estramustine (estradiol + mustine) shows a selective distribution to estrogen-dependent breast tumors (Fig. 4.9).

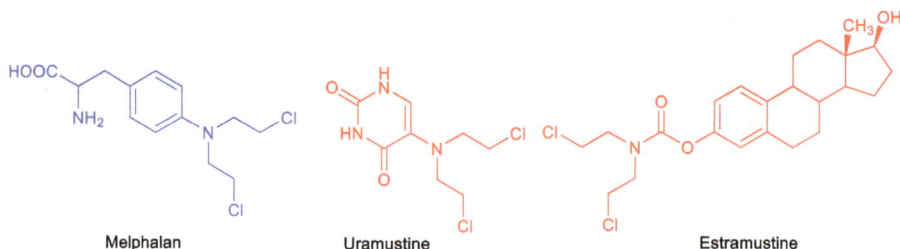

Fig. 4.9: Carrier groups increase cell or tissue selectivity of the drug.

3. Increased chemical stability, mainly against the acidic environment, to make it possible for the drug to be administered orally. Thus, by modifications in the side chain of penicillins, active compounds are obtained orally, with greater resistance to the acidic environment (Fig. 4.10).

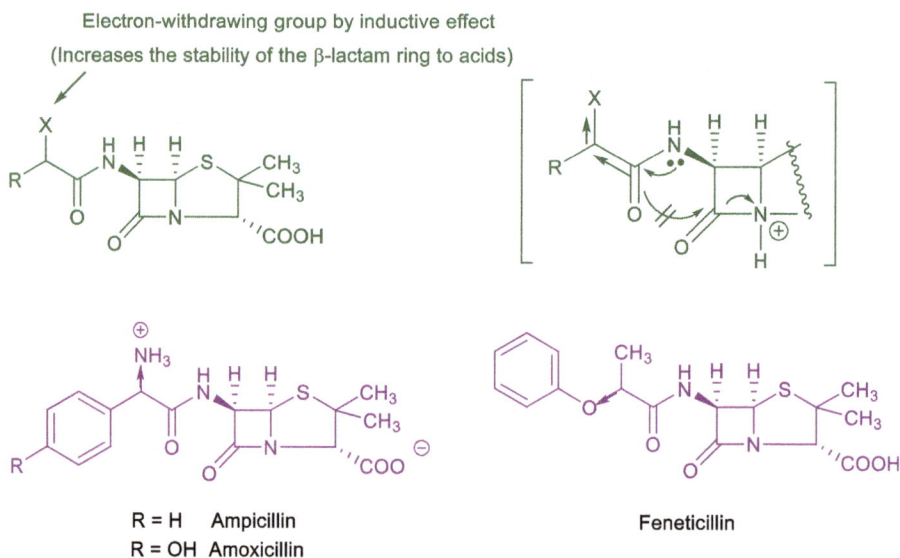

R = H   Ampicillin
R = OH   Amoxicillin

Feneticillin

Fig. 4.10: Increased chemical stability.

4.3 Strategies

There are three possible strategies: modulative, disjunctive, and conjunctive structural variations.

4.3.1 Modulative structural variation

This makes limited transformations in the structure of the model, which retains its fundamental structure:
- Vinylogy
- Homology
- Introduction of cyclic systems
- Introduction or substitution of polar groups, by nonpolar bulky ones
- Isosterism and bioisosterism

4.3.1.1 Vinylogy

It has been shown that certain molecules differing in one or more vinyl groups (CH = CH) located in the side chain or included in a cycle may have similarity in their pharmacological properties. Procaine vinylogous **A** and **B** are also local anesthetics. The hydrogenated compound **C**, homologous to **A** (a homolog of a given compound is the analog to it resulting from the addition (or subtraction) of one or more CH_2 to a chain or ring), lacks activity (Fig. 4.11).

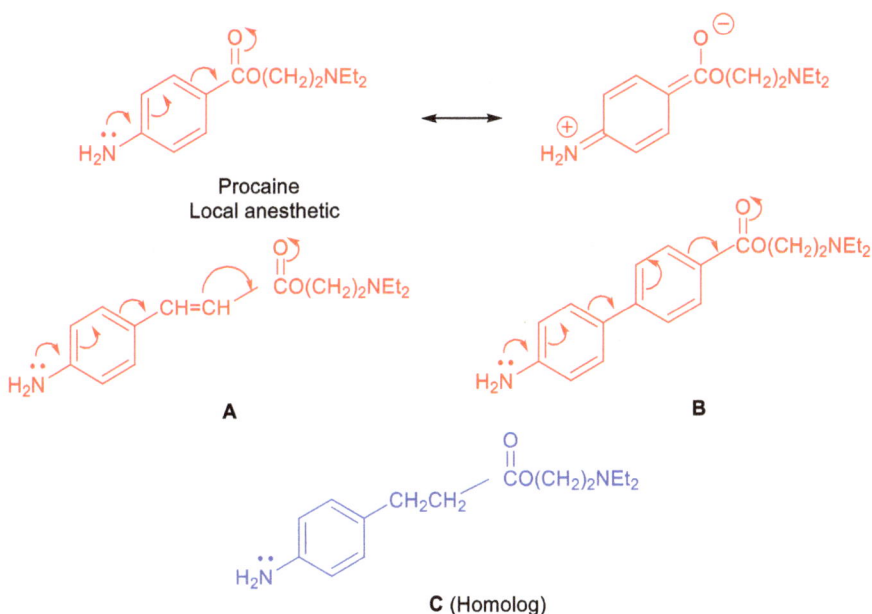

Fig. 4.11: Active vinylogous (**A** and **B**) of procaine; **C** is an inactive homolog of procaine.

- The resonant effect is not lessened by distance. The inductive effect does.
- If the electron density of a given area of a molecule is important for its binding to the receptor, the biological activity can be maintained in the reference vinylo-

gous. Sometimes, some surprising properties can be found in vinylogous products. Thus, the acetylcholine (AcC) vinilogous has nicotinic and no muscarinic activity, while AcC presents both (Fig. 4.12).

Fig. 4.12: Acetylcholine and vinylogous.

– Vinylogy serves to demonstrate if resonance effects are important in terms of activity (Fig. 4.13).

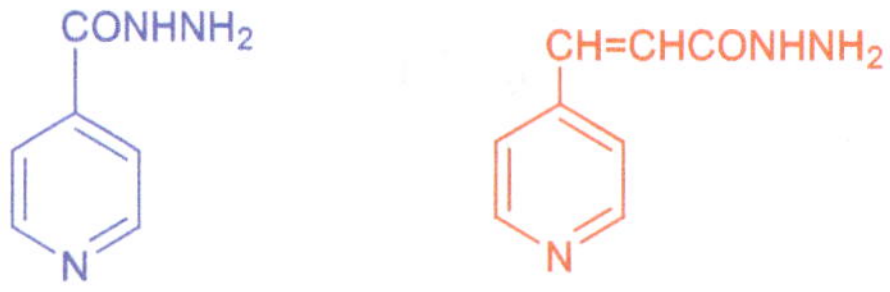

Fig. 4.13: Vinylogy might give rise to an inactive drug.

This example demonstrates that the activity lies in the distance that exists between the ring and the chain, which disappears when the activity increases, by introducing a vinyl group.

Vinylogy has also been applied in the opposite sense. Thus, removal of benzene ring from the sweetener dulcin gives rise to ethoxyurea, an equally sweet substance (Fig. 4.14). In short, vinylogy is the transmission of electronic effects through a conjugated bonding system.

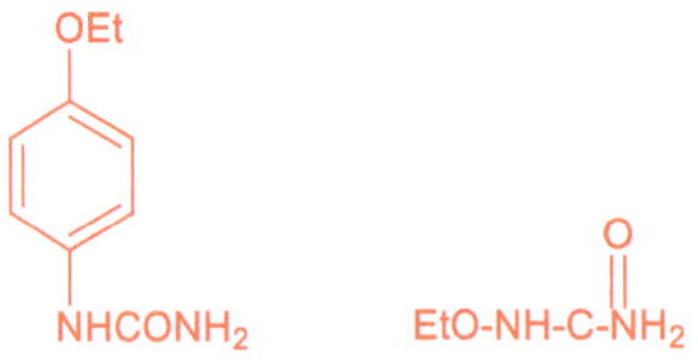

Fig. 4.14: An example in which elimination of a benzene ring maintains the sweetener characteristic.

4.3.1.2 Homology

Homologs are those molecules that differ from one another in a methylene group or more. Example: the anti-nicotinic drug hexamethonium (ganglioplejic effect, which blocks the sympathetic and parasympathetic glanglionic nerve transmission) and its decamethylene homologue decamethonium (curarizant effect, with a non-depolarizing skeletal muscle relaxant activity) (Fig. 4.15).

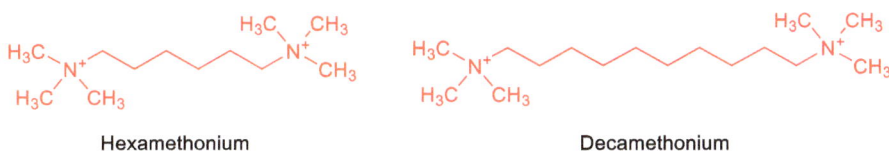

Hexamethonium   Decamethonium

Fig. 4.15: An example of homologous derivatives.

4.3.1.3 Introduction of cyclic systems

This approach is useful for the study of the active conformation in flexible molecules. The formation of a ring restricts the conformational freedom of the original molecule. This modification may also entail the creation of new stereocenters in the molecule. Examples: chlorpromazine and thioridazine (both phenothiazine neuroleptics), ondansetron (antiemetic), and cilansetron (an even more potent antiemetic) (Fig. 4.16).

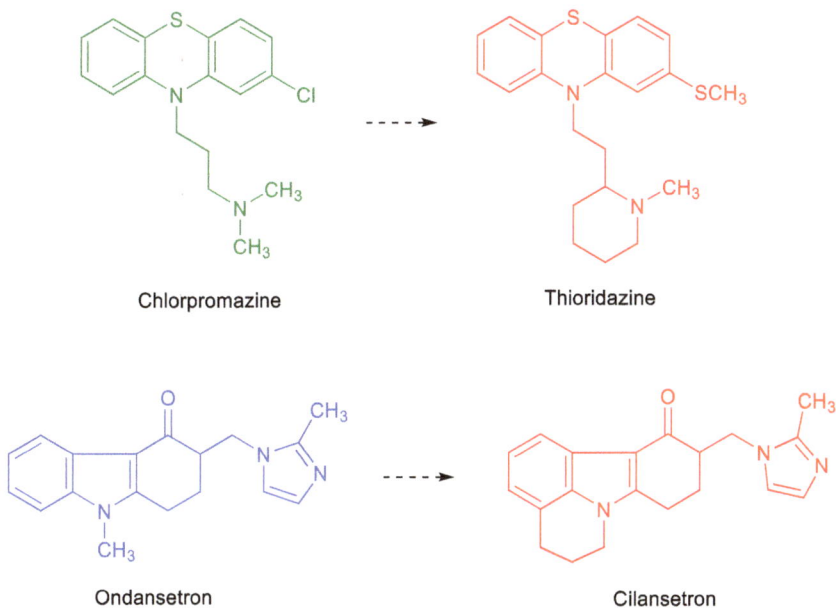

Chlorpromazine   Thioridazine

Ondansetron   Cilansetron

Fig. 4.16: Introduction of cycles.

4.3.1.4 Introduction or substitution of polar groups, by nonpolar bulky ones

This is a procedure of particular interest for converting agonists into antagonists. An antagonist is a substance having an action opposite to another to which it refers (Fig. 4.17).

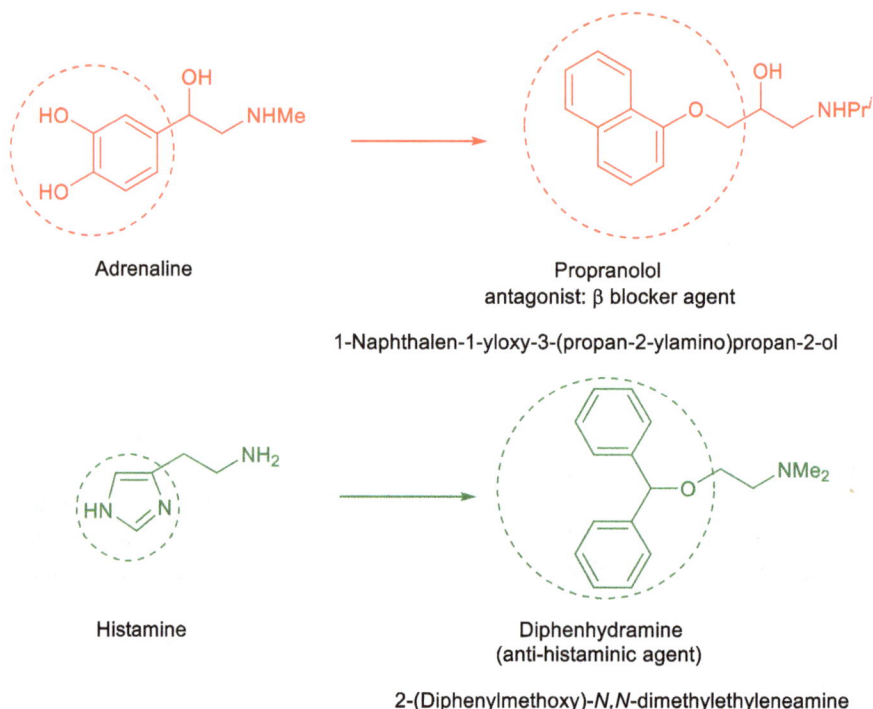

Adrenaline

Propranolol
antagonist: β blocker agent

1-Naphthalen-1-yloxy-3-(propan-2-ylamino)propan-2-ol

Histamine

Diphenhydramine
(anti-histaminic agent)

2-(Diphenylmethoxy)-*N*,*N*-dimethylethyleneamine

Fig. 4.17: Examples for the conversion of agonist into antagonist drugs.

Propranolol is used to treat tremors, angina (chest pain), hypertension, heart rhythm disorders, and other heart or circulatory conditions. It is also used to treat or prevent heart attacks.

Diphenhydramine is used to relieve red, irritated, itchy, watery eyes; sneezing; and runny nose caused by hay fever, allergies, or the common cold.

4.3.1.5 Isosterism and bioisosterism

4.3.1.5.1 Isosterism

One of the most frequent criteria for molecular variation is that of isosterism, which essentially consists of substituting atoms or groups of atoms equivalent in size and electronic distribution. It is a chemical concept:

- Langmuir concept (1919)
- Grimm's hydride displacement law (1925)
- Peripheral isoelectricity (Erlenmeyer, 1932)

(1) Langmuir defined isosteres as molecules or groups of atoms with the same number of atoms and valence electrons (Fig. 4.18).

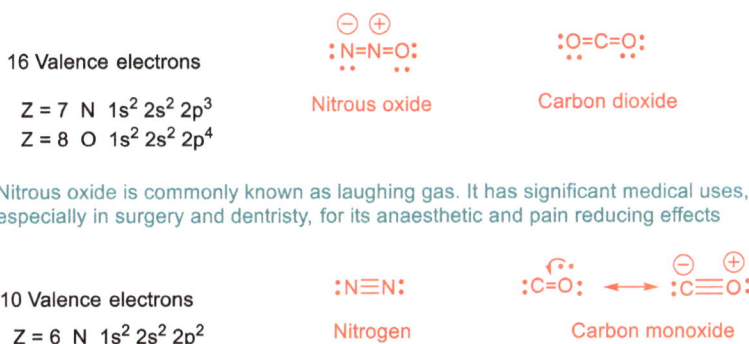

16 Valence electrons

$Z = 7$ N $1s^2\, 2s^2\, 2p^3$
$Z = 8$ O $1s^2\, 2s^2\, 2p^4$

:N=N=O:
Nitrous oxide

:O=C=O:
Carbon dioxide

Nitrous oxide is commonly known as laughing gas. It has significant medical uses, especially in surgery and dentristy, for its anaesthetic and pain reducing effects

10 Valence electrons

$Z = 6$ N $1s^2\, 2s^2\, 2p^2$

:N≡N:
Nitrogen

:C=O: ⟷ :C≡O:
Carbon monoxide

Fig. 4.18: Examples of isosteric molecules.

(2) Grimm formulated the so-called hydride displacement law, according to which the addition of a hydrogen atom to an atom of atomic number "n" provides a species with the same properties of the atom of higher atomic number "$n + 1$" (Fig. 4.19). In this way, families of "pseudo-atoms" with common electronic characteristics arose.

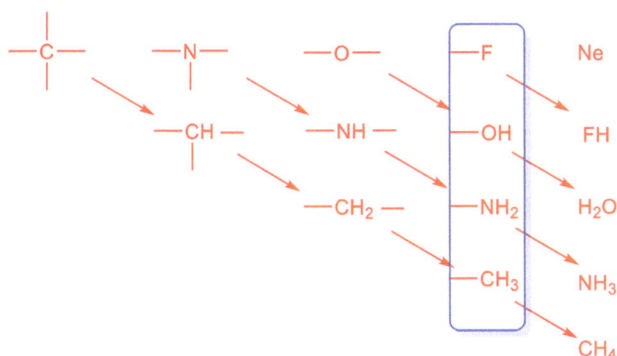

Fig. 4.19: Grimm's hydride displacement law. In each vertical column, the atom is followed by its pseudoatom.

(3) In 1932, Erlenmeyer extended the concept of isosterism by proposing a broader definition of isosteres as those elements, molecules, or ions with a similar electronic distribution in their valence layer. The concept of isosterism was thus extended to the elements of a column of the periodic system (e.g. carbon would be isostere of silicon and oxygen of sulfur). In this way, the so-called annular equivalents were defined (clusters that can be interchanged in a ring without giving rise to a substantial change in the physical and chemical properties of the latter): the following groups $-CH = CH-$ and $-S-$ are examples of equivalent ring groups that permit the explanation of the analogies between benzene and thiophene. The above criteria lead to a series of "ring equivalents", widely used for the design of analogs (Fig. 4.20).

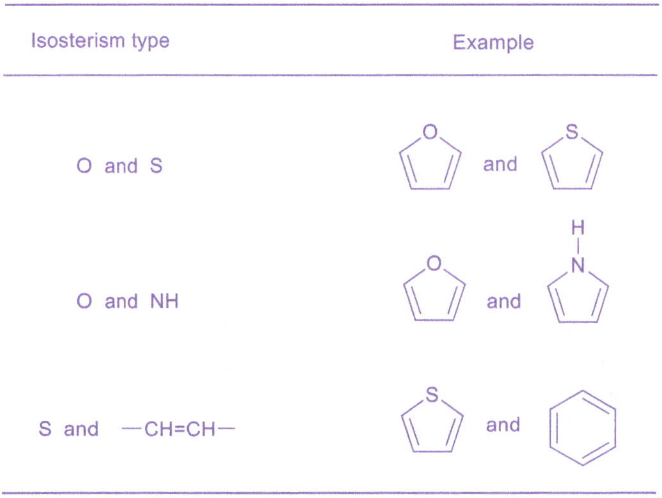

Isosterism type	Example
O and S	and
O and NH	and
S and —CH=CH—	and

Fig. 4.20: Isosteric equivalences between some aromatic rings.

Cyclo-equivalents are the atoms or groups of atoms that can be replaced in cycles without having a marked variation in their properties. All of the above are classical isosteres.

Friedman in 1951 introduced the term "bioisosterism" (a biological concept) to designate those molecules or atomic groups responsible for the same biological (or antagonistic) activity as a result of a similarity in their physical and chemical properties. Thus, the "nonclassical isostere" arises corresponding to those groups of atoms that, when introduced into a given molecule, generate compounds whose shape, size, or other properties make them bioequivalent, despite not being isoelectronic. The term "bioisosterism" acquires a broader character than the strict chemical notion mentioned before. It is important to keep in mind that the term "bioisostere" will be meaningless if the property or properties that make it comparable to a pair of chemical species are not indicated. Thus, for example, two classical isosteres according to

Grimm's law such as –OH and –CH$_3$ can be considered as bioisosteres in volume but not in terms of their lipophilia or in their electronic distribution.

As an example, sulfur atom is approximately equivalent to the vinylene group: Compare the boiling points of several benzene and thiophene derivatives (Fig. 4.21).

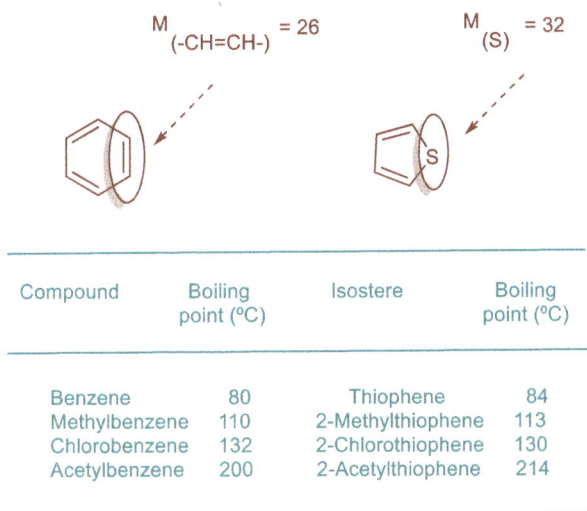

M$_{(-CH=CH-)}$ = 26   M$_{(S)}$ = 32

Compound	Boiling point (°C)	Isostere	Boiling point (°C)
Benzene	80	Thiophene	84
Methylbenzene	110	2-Methylthiophene	113
Chlorobenzene	132	2-Chlorothiophene	130
Acetylbenzene	200	2-Acetylthiophene	214

Fig. 4.21: Similarity of boiling points among several benzene and thiophene derivatives.

All the above are classical isosteres. In the concept of isosterism, there is an evolution: the first concept (Langmuir) is very restrictive, and gradually, a broader vision is given.

4.3.1.5.2 Bioisosterism

It is possible to indicate the following properties:

(a) **Size:** H, F (see Chapter 13 of Volume 2 of this series)

Br, Pr^i

I, Bu^t

Fluorine atom is considerably smaller than the other halogens. The fluorine derivatives differ from the rest of the halogen derivatives in which the fluorine atom forms particularly stable carbon bonds and, unlike what happens to the other halogens, is very rarely ionized or displaced. Therefore, due to both its chemical inertia and its small size, fluorine is often compared to hydrogen.

(b) **Electronic distribution and polarizability** (Fig. 4.22)

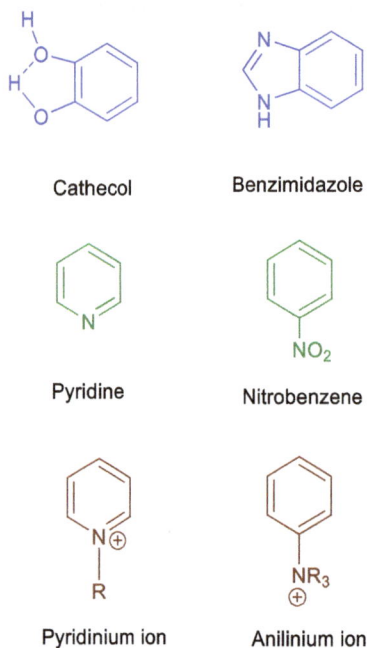

Cathecol   Benzimidazole

Pyridine   Nitrobenzene

Pyridinium ion   Anilinium ion

Fig. 4.22: Bioisosteres based on analogous electronic distribution and polarizability.

Catechol is a bioisostere of benzimidazole in the sense that while in the first case a second ring may be formed through an intramolecular bond, in the second case, the imidazole ring mimics the "five-membered ring" of catechol through a cycle formed with covalent bonds.

In the same way, the bioisosterism between thiourea, N-cyanoguanidine, and β-nitroketene aminal groups (Fig. 4.23) has been demonstrated in histamine H_2-receptor antagonists, also known as H2-blockers (burimamide, cimetidine, and ranitidine). We will go more deep into this aspect in Chapter 7 of Volume 2 of this series.

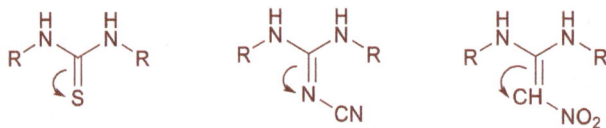

Fig. 4.23: Bioisosterism between the thiourea, N-cyanoguanidino, and β-nitroketene aminal groups.

When the biological activity depends fundamentally on the distribution of charges and polar effects of the molecule, the maintenance of these factors can lead to bioisosteric equivalences between very different molecules. Pyridine and nitrobenzene may serve to illustrate this possibility. To understand this relationship, we compare the structures of benzene with that of pyridine and those of pentagonal aromatic hetero-cycles, such as furan, thiophene, and pyrrole (Scheme 4.1).

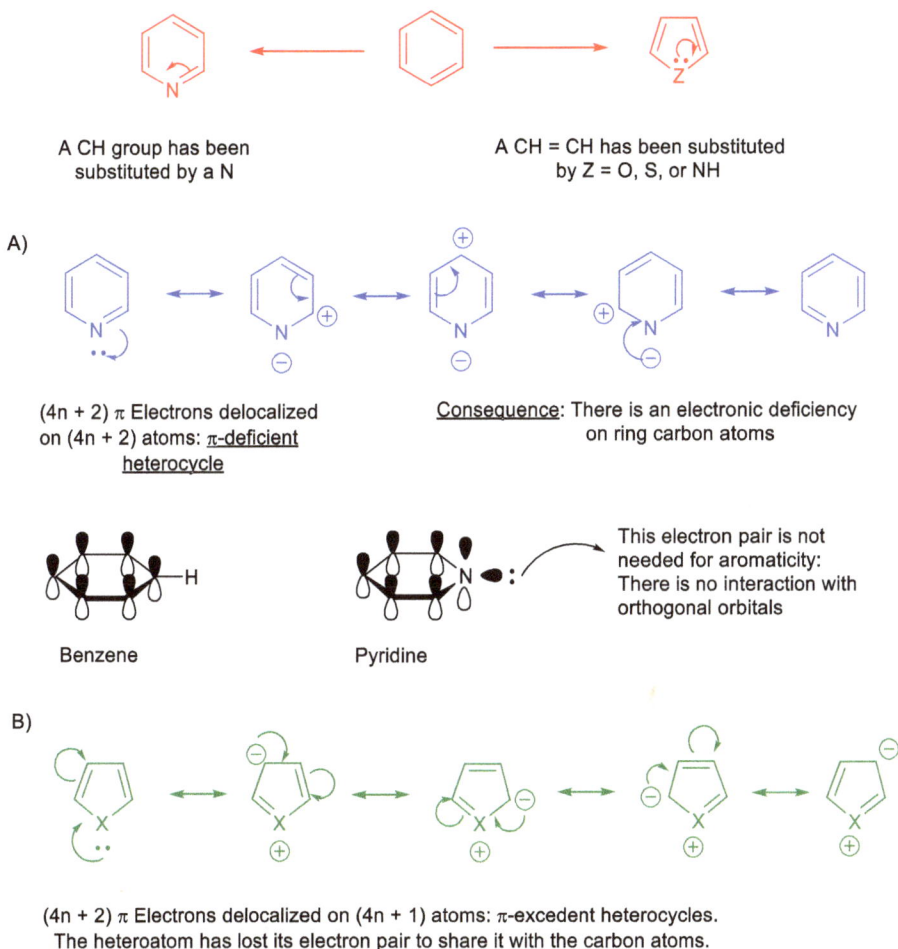

A CH group has been substituted by a N

A CH = CH has been substituted by Z = O, S, or NH

A)

(4n + 2) π Electrons delocalized on (4n + 2) atoms: π-deficient heterocycle

Consequence: There is an electronic deficiency on ring carbon atoms

Benzene

Pyridine

This electron pair is not needed for aromaticity: There is no interaction with orthogonal orbitals

B)

(4n + 2) π Electrons delocalized on (4n + 1) atoms: π-excedent heterocycles. The heteroatom has lost its electron pair to share it with the carbon atoms.

Scheme 4.1: A π-deficient aromatic heterocycle (pyridine) and π-excedent aromatic heterocycles (furan, thiophene, and pyrrole).

The dipole moment helps in explaining the distribution of charge of a molecule: Dipole moment expresses the polar character or polarity of molecules. The product of charge and distance of separation of atoms in a chemical bond defines the term "dipole moment of molecules". If $+q$ is the amount of positive charge separated by $-q$ amount of negative charge by the bond distance l, then dipole moment (μ) of the molecule = $q \times l$. The dipole moment allows us to verify experimentally the distribution of charges (Fig. 4.24).

There would be three possibilities:

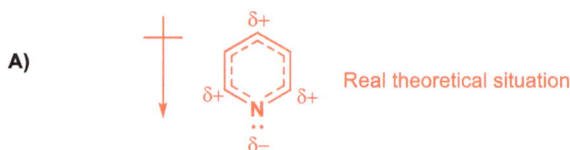

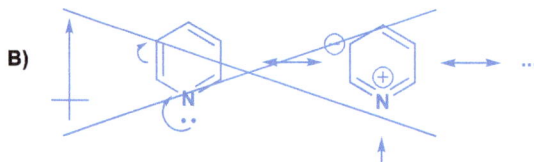

A)

Real theoretical situation

B)

Geometrically it is not possible,
because the strain would be enormous

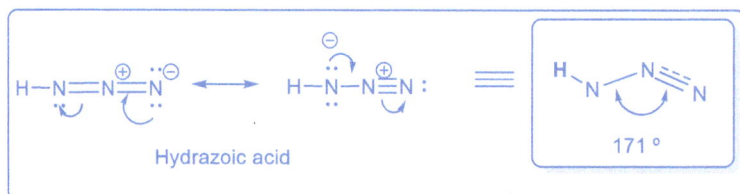

Hydrazoic acid

171 °

A regular hexagon is a hexagon whose sides and angles measure the same.
The (interior) angles measure 120°, whereas in hydrazoic acid the angle formed
by the intermediate N atom with the other bonds that hold it to two other N atoms
is 171°. Therefore, the situation in **B)** is highly unstable, from the geometrical point
of view.

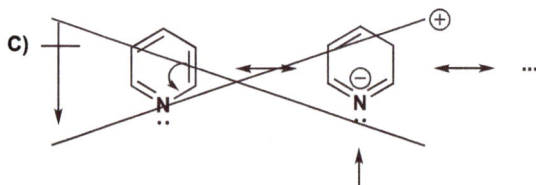

C)

Valence-shell occupancy exceedes 8 for N, and as a result, it is not possible

Fig. 4.24: Several possibilities for the electronic distribution of pyridine, although only one is the real one.

The consequences are very evident: pyridine will react fundamentally according to an
S$_N$Ar (nucleophilic aromatic substitution), while the three aromatic pentagonal heterocycles will do so according to the S$_E$Ar (aromatic electrophilic substitution). Chichibabin reaction is a method for producing 2-aminopyridine derivatives by the reaction
of pyridine with sodium amide (Fig. 4.25). The mechanism of the formation of 2-
aminopyridine by the Chichibabin reaction will be explained in Chapter 7 of Volume 2
of this series.

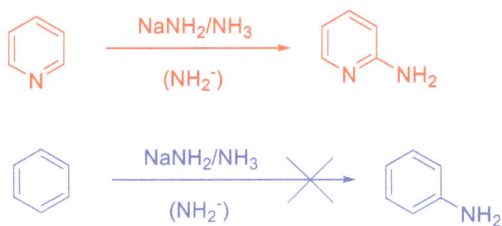

Fig. 4.25: Reactivity of pyridine against $S_N Ar$.

Pyridine and nitrobenzene resist $S_E Ar$, and when they do so under these conditions, they direct the entrance of the electrophilic groups (E^+) toward the *meta* position. This suggests a certain polar similarity, and consequently, the possibility of bioequivalence: *N,N*-diethyl-*m*-nitrobenzamide, for example, possesses analeptic properties (an analeptic agent is a stimulating drug of the central nervous system) very similar to those of its pyridine analog, nikethamide (Fig. 4.26).

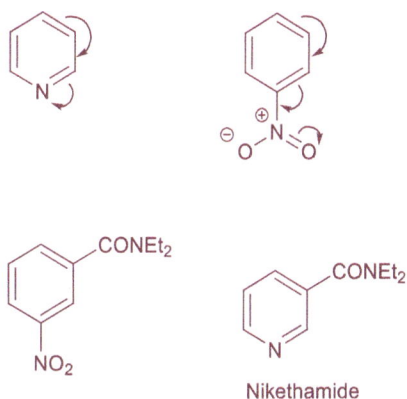

Nikethamide

Fig. 4.26: *N,N*-Diethyl-*m*-nitrobenzamide and nikethamide possess similar analeptic properties.

(c) Solubility in lipids, as in the –CH_2– and –S– groups, or in the trimethylene and *p*-phenylene groups (Fig. 4.27).

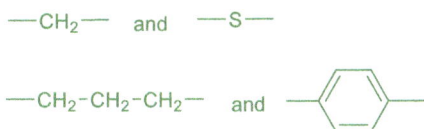

Fig. 4.27: Groups with the same solubility in lipids.

(d) pK_a, as in the acidic groups in Fig. 4.28.

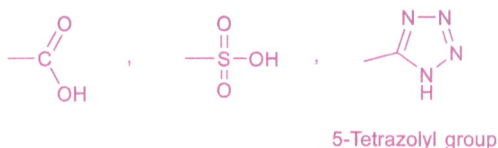

5-Tetrazolyl group

Fig. 4.28: Groups with similar acidity.

In principle, the acidic character of the 5-tetrazolyl moiety and consequently its bioisosteric relationship with the carboxylic acid group (Scheme 4.2) may be surprising.

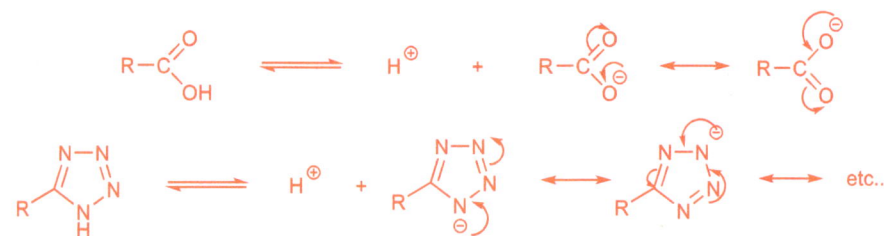

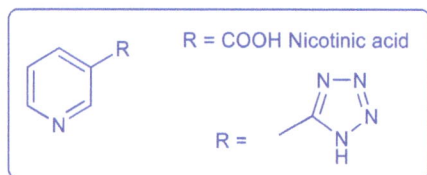

R = COOH Nicotinic acid

R =

Looking for new and better antihyperlipidemic agents, the tetrazolyl analogue of nicotinic acid turned out to be also so active, lowering the level of cholesterol in the blood.

Scheme 4.2: Acidic characters of the carboxyl and of the 5-tetrazolyl groups.

While the acidity of the carboxylic acid (p$K_a \approx 4.2$–4.4) is related to the resonance stabilization of the carboxylate anion, the acidity of tetrazole (p$K_a \approx 4.9$) is attributed to the delocalization of the negative charge on each of the nitrogen atoms of the five-membered ring. Although the greater number of resonant forms of tetrazole, which contribute to the final hybrid, might suggest that the resonance energy of tetrazole was more effective than that of the carboxylate anion, the higher electronegativity of the oxygen atom relative to that of nitrogen is a decisive factor to justify the higher acidity of the carboxylic acid group.

(e) Ability to establish hydrogen bonds, as in the –OH group of phenol, –NH of CH$_3$SO$_2$ NH–, and R-NH-CO-NH–.

4.4 Disjunctive replication

This means reducing the structure of the model until you keep nothing more than the essential part. These analogs are partial replicates of the prototype drug. The technique of opening cycles is used. Disjunction: action and effect of separating.

(a) Procaine is the result of the simplification of cocaine (Fig. 4.29).

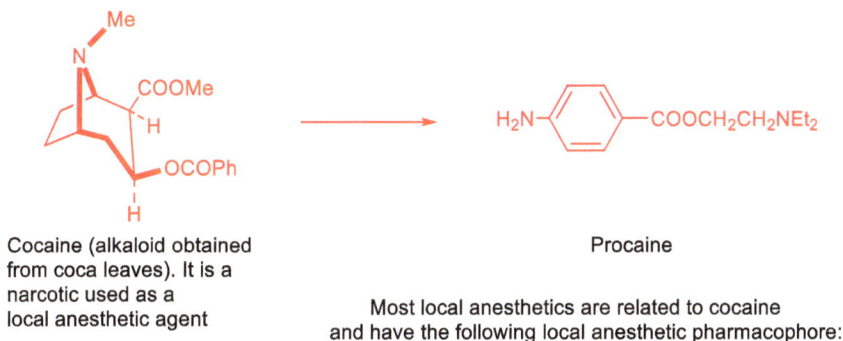

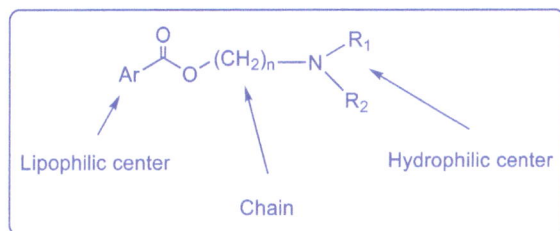

Cocaine (alkaloid obtained from coca leaves). It is a narcotic used as a local anesthetic agent

Procaine

Most local anesthetics are related to cocaine and have the following local anesthetic pharmacophore:

Lipophilic center

Hydrophilic center

Chain

Fig. 4.29: Simplification of the prototype (disjunctive replication) cocaine.

(b) Some hypnoanalgesics derived from the simplification of morphine (Fig. 4.30). From the comparison of these structures, it can be deduced that the analgesic activity of morphine is associated with the presence of a benzene ring attached to a quaternary carbon and this to a tertiary amine through a two-carbon chain; this fragment is the pharmacophore of hypnoanalgesic drugs (Fig. 4.30).

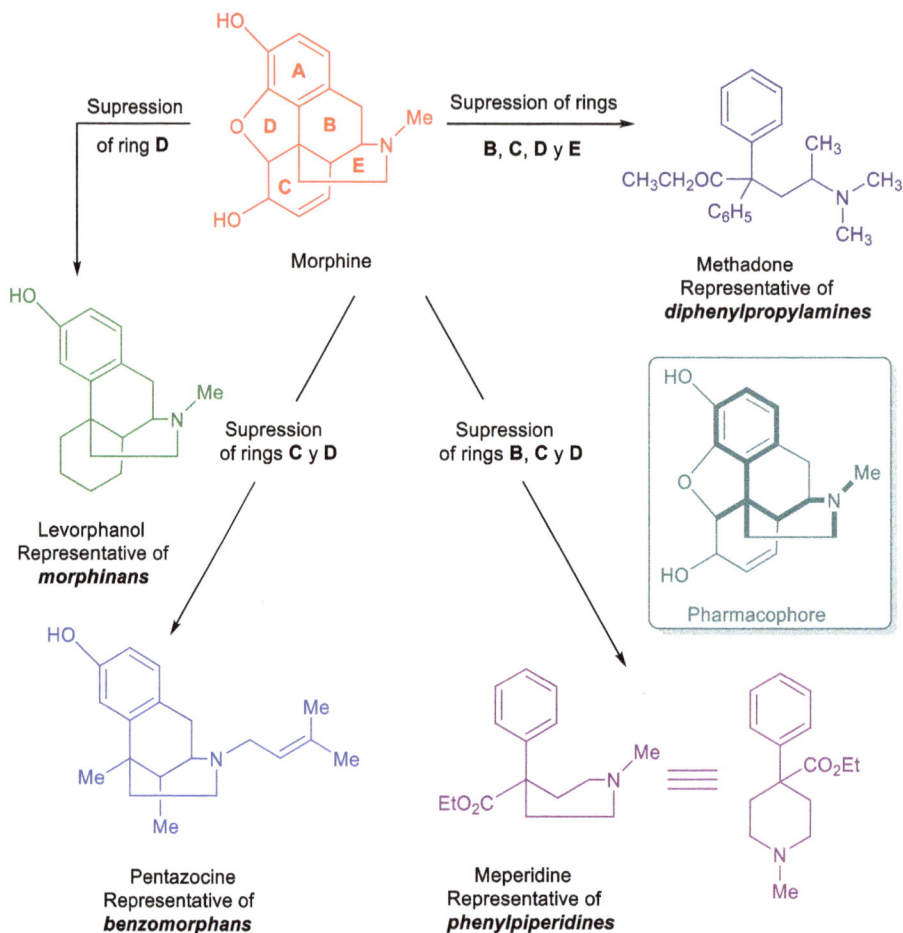

Fig. 4.30: Simplification of the prototype (disjunctive replication) morphine.

4.5 Conjunctive replication

Larger replicas than the model are made. It has two aspects:

– Molecular duplication (of interest in the development of antimetabolites, Fig. 4.31).

Antimetabolites have a structure very similar to that of the natural substrate, resulting in biologically unusable products through the incorporation of molecules other than the natural ones.

– Hybrid or molecular combination. This involves the combination of two different molecules (Fig. 4.32).

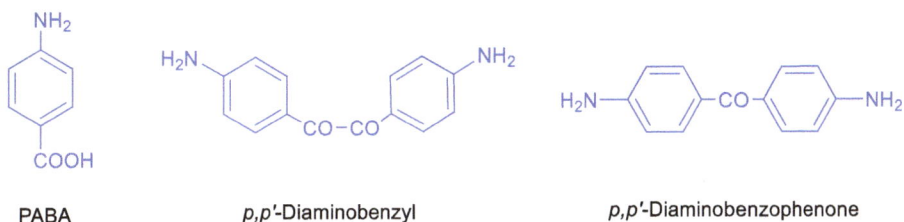

Fig. 4.31: Molecular duplication of the prototype PABA (conjunctive replication).

PABA p,p'-Diaminobenzyl p,p'-Diaminobenzophenone

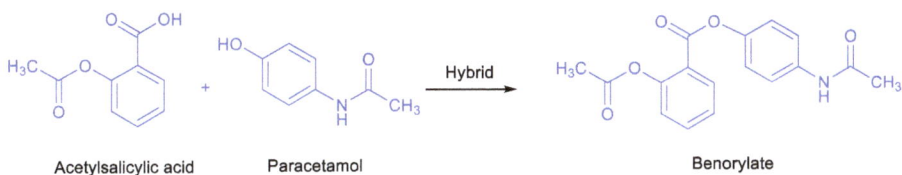

Acetylsalicylic acid Paracetamol Benorylate

Fig. 4.32: Molecular combination of the prototypes (disjunctive replication) acetylsalicylic acid and paracetamol.

4.6 Peptidomimetics

Biologically active molecules containing amide bonds normally suffer from pharmacokinetic hazards. Bioisosteric transformations have been carried out in the carboxamide group in order to increase its stability with great success in the area of peptidomimetics (these can be defined as structures capable of replacing peptides in their interactions with receptors and enzymes). The most established modifications are as follows: *N*-methylation, change of configuration (configuration D), formation of a retroamide or azapeptide, use of aminoisobutyric acid or dehydroamino acids, substitution of the amide bond by an ester (depsipeptides), introduction of the ketomethylene group, introduction of the thioamide group, reduction of the amide carbonyl group, and use of an olefinic double bond (Fig. 4.33).

The best known peptidomimetic is morphine, which owes its analgesic action to its binding capacity to opioid receptors such as the endogenous peptides, enkephalins and endorphins (Fig. 4.34).

4.7 Summary

The design and development of a lead compound into a drug is a laborious and often costly process, with most candidates failing due to metabolism and pharmacokinetic issues rather than potency. Bioisosteric replacement is a strategy used by medicinal chemists to address these limitations while retaining the potency/efficacy of the initial

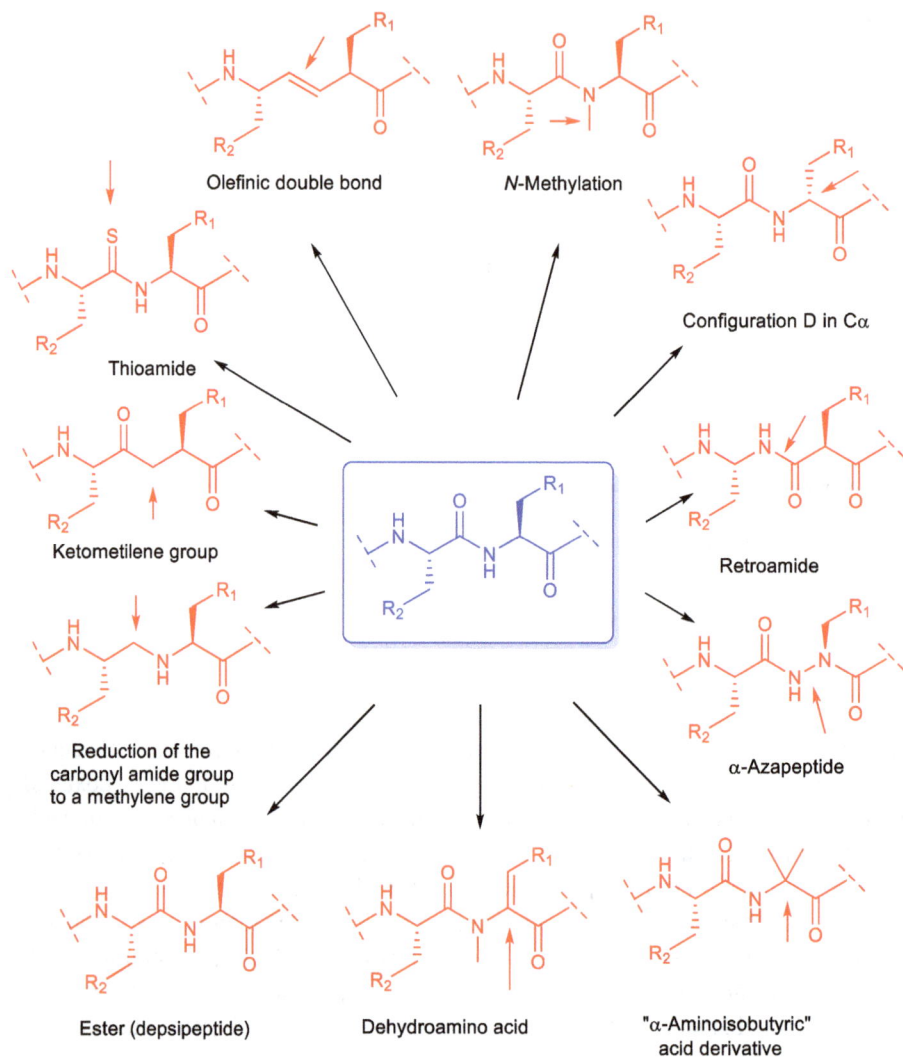

Fig. 4.33: Isosteric substitutions of the peptide bond.

lead compound. The use of bioisosteres and the introduction of structural changes to the lead compound allows the chemist to alter the compound's size, shape, electronic distribution, polarizability, dipole, polarity, lipophilicity, and pK_a, while retaining potent target engagement. Therefore, the bioisosteric approach can be used for the rational modification of a lead compound toward a more attractive therapeutic agent with improved potency, selectivity, altered physical, metabolic, toxicological properties with the bonus of generating novel intellectual property.

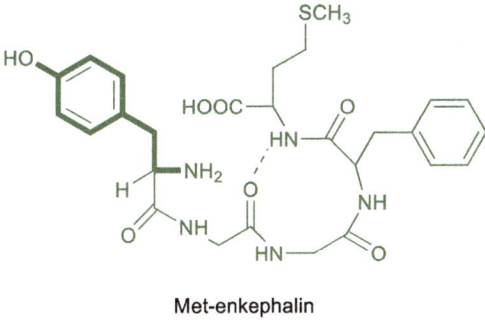

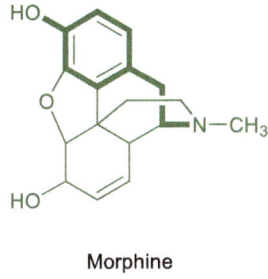

Met-enkephalin   **Morphine**

Fig. 4.34: Met-enkephalin and morphine.

4.8 Exercises

1. Indicate the molecular modification criteria used in the design of the following drugs, giving a reasoned explanation:

(a)

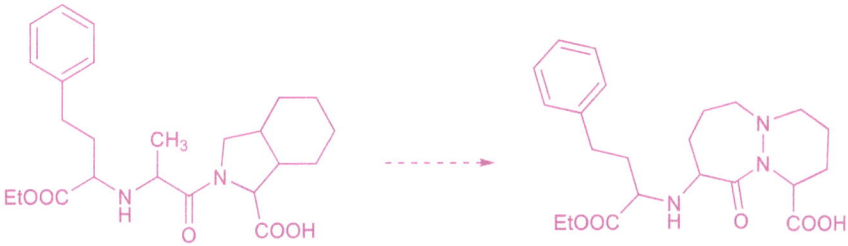

(b)

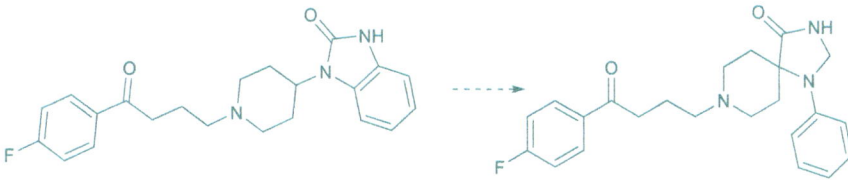

(c)

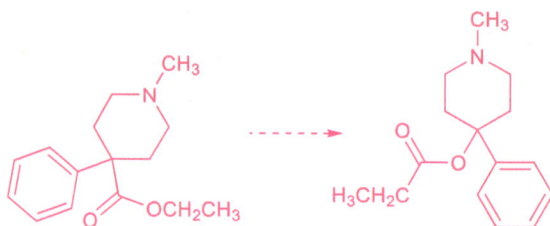

(d)

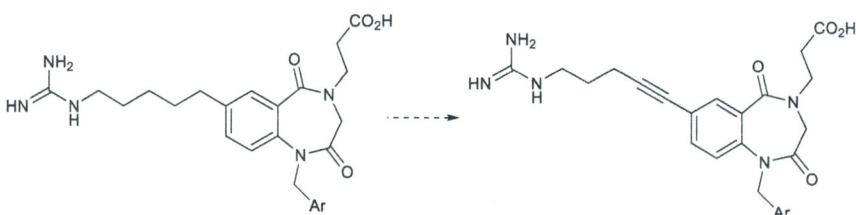

(e)

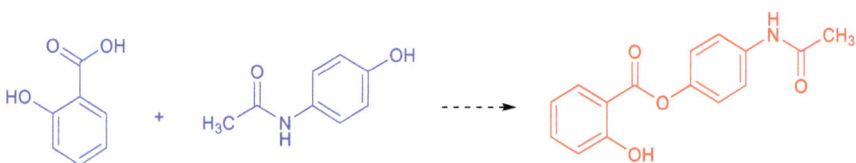

(f)

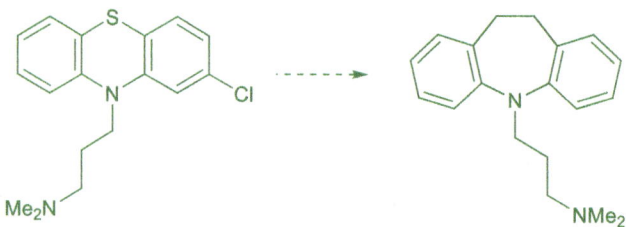

(g)

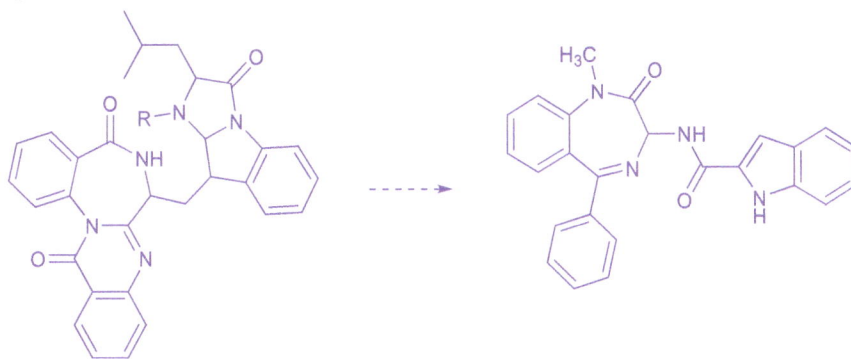

(h)

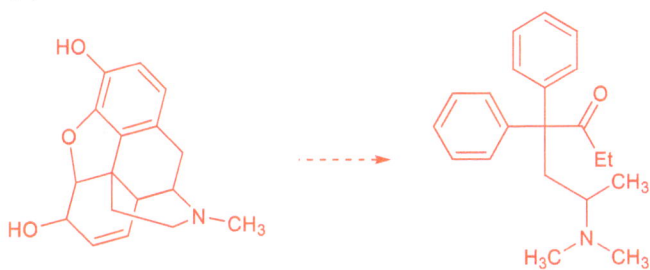

(i)

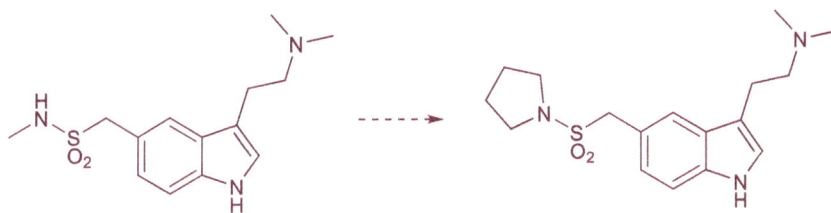

(j)

(k)

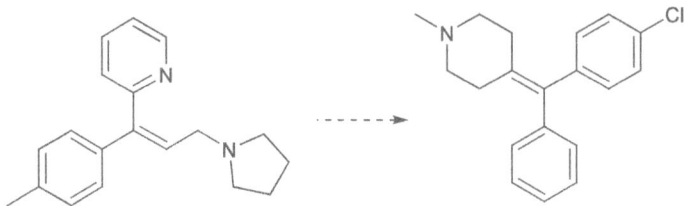

(l)

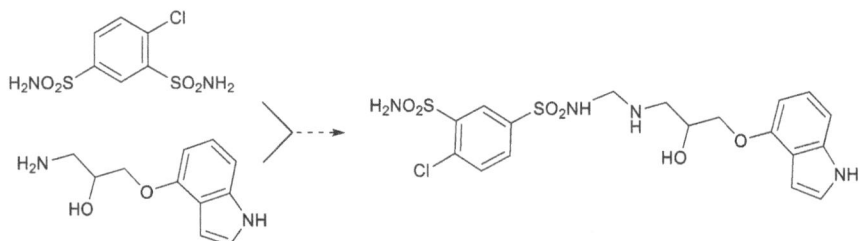

(m)

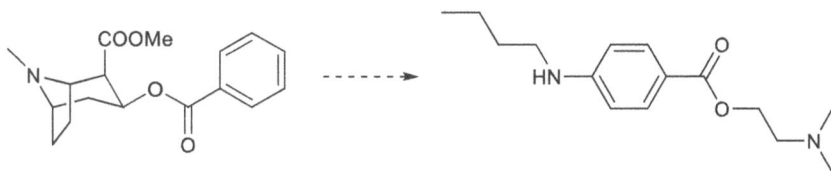

(n)

2. Given the following pairs of structures, indicate if they are isosteres, homologues, or vinylogues:

(a)

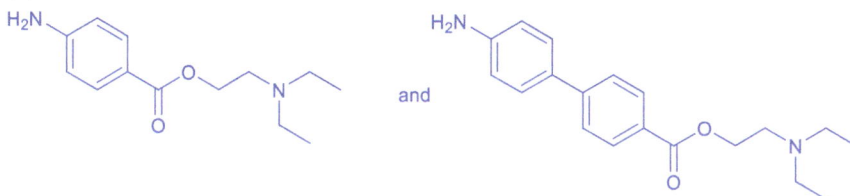

and

(b)

and

(c)

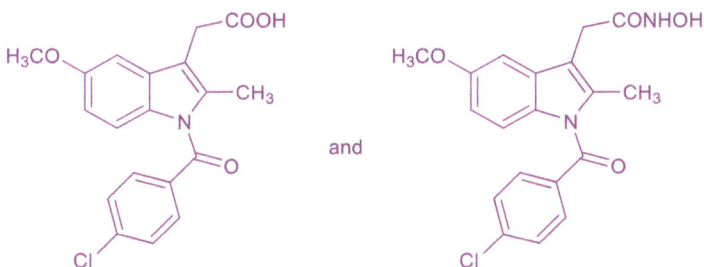

and

(d)

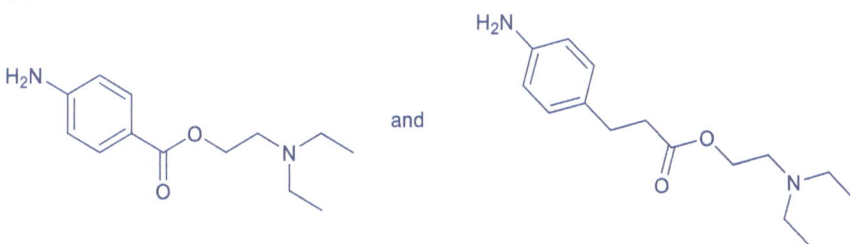

and

(e)

and

(f)

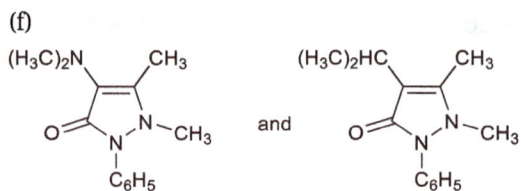

and

(g)

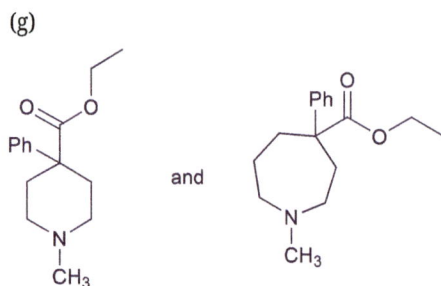

and

(h)

and

(i)

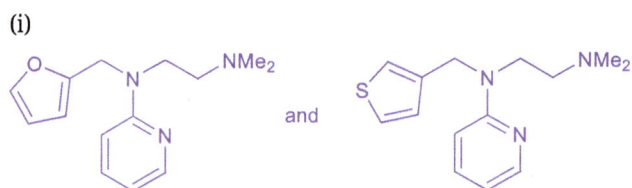

and

(j)

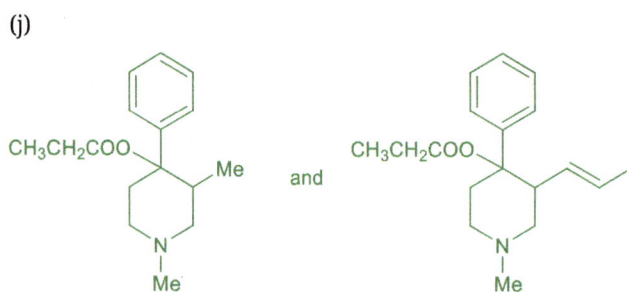

and

(k)

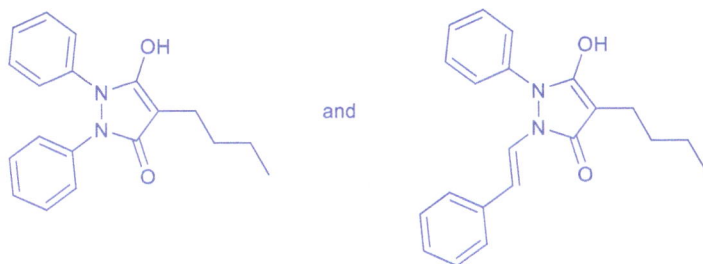

and

3. Explain the purpose of the molecular modification achieved when switching from one drug to the other:

(a)

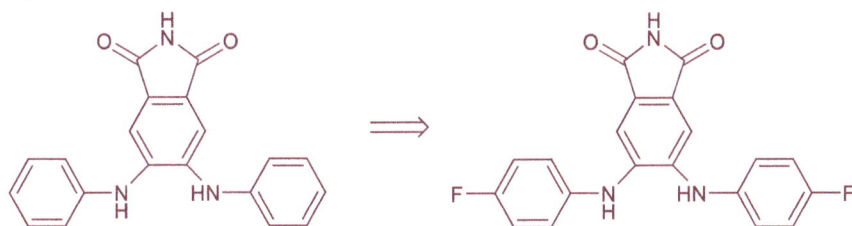

(b)

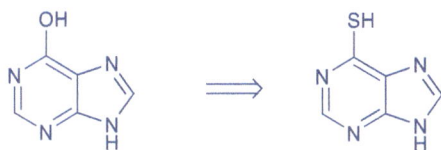

(c)

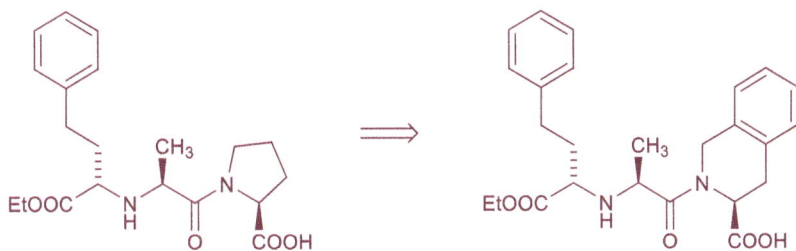

(d)

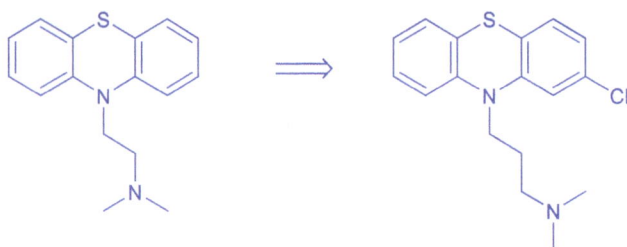

(e)

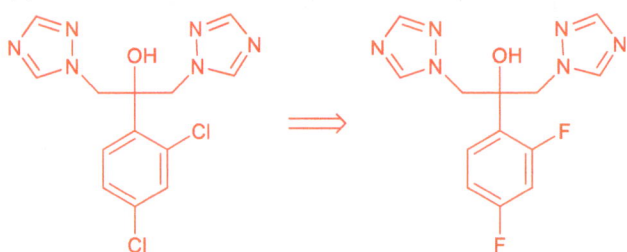

(f)

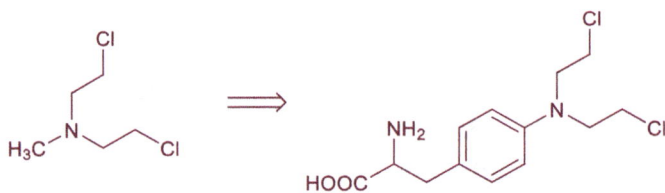

(g)

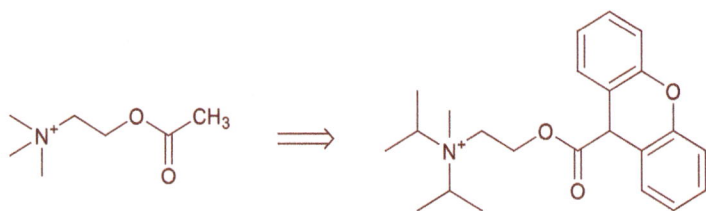

(h)

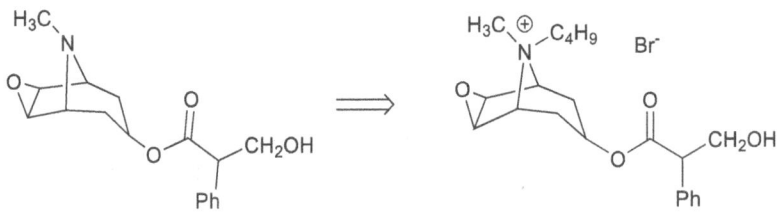

5 Biological targets and receptors for drugs

5.1 Goals

- To understand the concept of biological target and natural ligand
- To know the interactions that allow drug–receptor molecular recognition as a cause of chemical affinity, efficacy, or intrinsic activity
- To know stereochemical requirements of the drugs
- To know the existence of drugs that do not bind at the receptor-binding site

5.2 Membrane receptor

A membrane receptor is a macromolecule embedded in the cell membrane, with one part of its structure oriented toward the outside and the other toward the interior of the cell.

There are a variety of chemical messengers or neurotransmitters that interact with receptors: some are simple molecules, such as a quaternary ammonium salt (acetylcholine), monoamines (noradrenaline, dopamine, and serotonin), or amino acids (e.g. γ-aminobutyric acid, glutamic acid, and glycine) (Fig. 5.1). Other chemical messen-

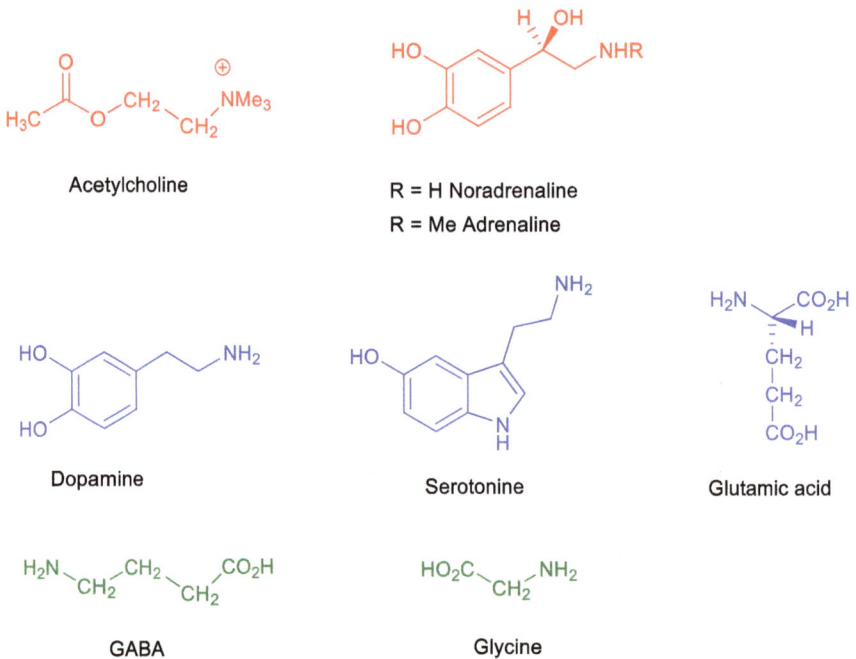

Acetylcholine

R = H Noradrenaline
R = Me Adrenaline

Dopamine

Serotonine

Glutamic acid

GABA

Glycine

Fig. 5.1: Examples of neurotransmitters and the hormone adrenaline.

https://doi.org/10.1515/9783111316901-005

gers are more complex, and include lipids such as prostaglandins, neuropeptides such as endorphins and enkephalins, and peptide hormones such as angiotensin.

5.3 Affinity and intrinsic activity: agonists and antagonists

There are two parameters defining the behavior of the drug in its relationship with receptors:
1. Its ability to bind to receptors or affinity.
2. The ability to activate them, intrinsic activity or effectiveness.

According to the presence and absence of the described properties, drugs are classified into:
−   Agonists: substances that possess receptor affinity and intrinsic activity.
−   Antagonists: drugs that possess affinity, but not intrinsic activity, so they occupy the receptor but are unable to activate it. In addition, they will oppose its occupation by agonists and will prevent it from producing an effect:

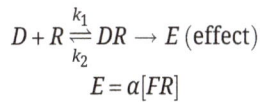

$$D + R \underset{k_2}{\overset{k_1}{\rightleftharpoons}} DR \rightarrow E \text{ (effect)}$$
$$E = \alpha[FR]$$

Pure agonist: $\alpha = 1$ Pure *antagonist*: $\alpha = 0$
Partial agonist: $0 < \alpha < 1$

The receptor may be
−   linked to an ion channel;
−   have catalytic capacity;
−   associated with a G protein;
−   intracellular.

5.4 Types of receptors

5.4.1 Ionic channels

The membrane is formed by a bilayer of molecules, whereby the center of the cell membrane is hydrophobic (represented by wavy lines in Fig. 5.2), whereas the termini, which are directed toward the outside and to the cytoplasm of the cell, are hydrophilic (represented by circles in Fig. 5.2). This characteristic makes it difficult for polar molecules and ions to enter or leave the cell, and the movement of sodium and potassium ions through the membrane is crucial for the functioning of the nerves.

Ionic channels are complexes formed by five protein subunits that cross the cell membrane. The center of the complex is hollow and is furrowed by polar amino acids, which translates into a hydrophilic tunnel. Ions can traverse the fat barrier of the cell membrane, moving through these hydrophilic channels or tunnels. However, there has to be some kind of control: in other words, there should be a lock that opens or closes as required. In the idle state, the ion channel remains closed. However, when a chemical messenger binds to the outer binding site of the receptor protein (one part of which is the ion channel), an induced adjustment occurs which causes the protein to change its shape, the channel to open, allowing the passage of the ions through the channel.

Ionic channels are specific for certain ions. For example, there are different cationic channels for Na^+, K^+, and Ca^{2+} ions. There are also anionic channels for the Cl^- anion.

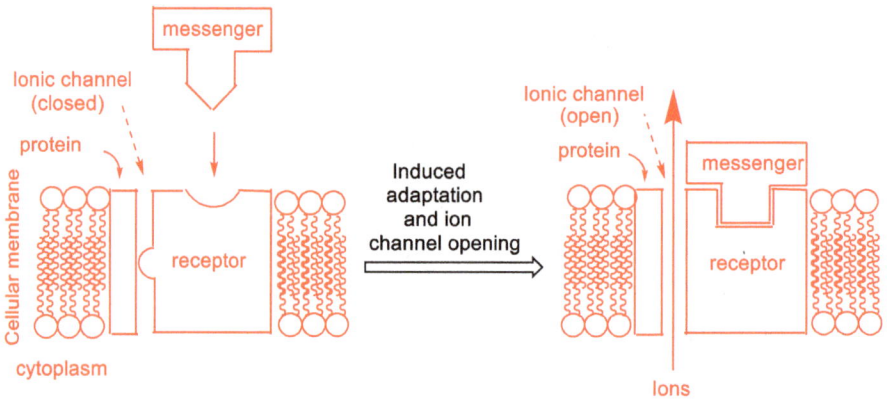

Fig. 5.2: Structure of an ionic channel.

Example: Nicotinic receptor of neurotransmitter acetylcholine.

5.4.2 Receptors with an intrinsic catalytic activity

The binding of a ligand on the extracellular domain of the receptor triggers the activation of an enzymatic system located in the intracellular domain of such a receptor (Fig. 5.3).

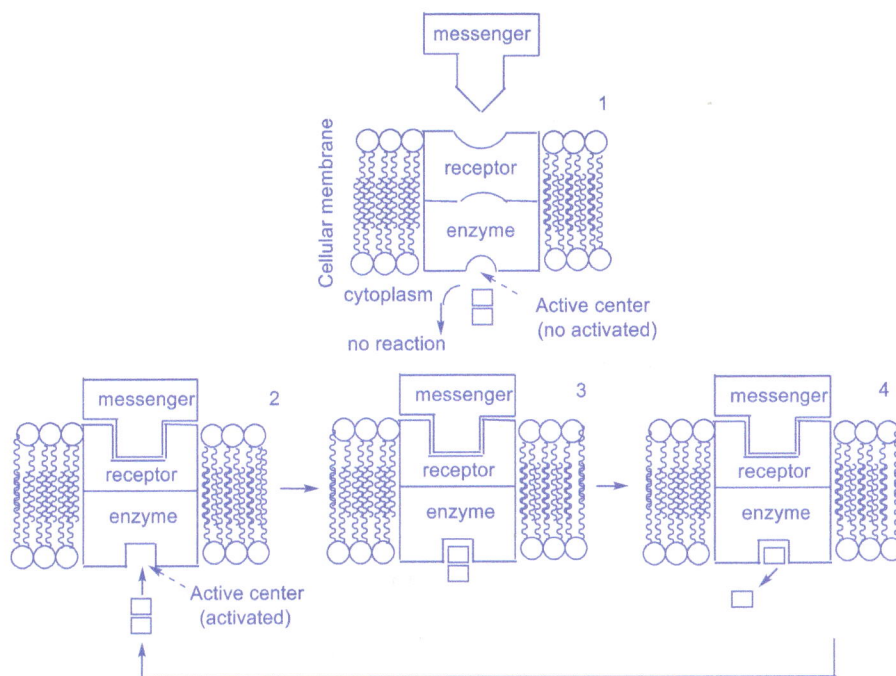

Fig. 5.3: Mechanism of activation of an enzymatic system associated with a membrane receptor: (1) rest state; (2) the enzyme acquires a productive conformation by interaction of the messenger with the receptor; (3) and (4): enzymatic process and return to (2).

Example: Certain protein kinases.

5.4.3 Receptors bound to G proteins: metabotropic receptors

5.4.3.1 Guanosine diphosphate

Guanosine diphosphate (GDP) is a nucleotide diphosphate. It is an ester of pyrophos-phoric acid with the nucleoside guanosine. GDP consists of a pyrophosphate group, a pentose sugar (D-ribose), and the nucleobase guanine (Fig. 5.4).

Receptors bound to G proteins lead to the formation of secondary messengers that are responsible for biochemical responses. They consist of the binding of the L-R complex with a G protein (so called because it is associated with the guanine nucleotide, GDP). G proteins are composed of three subunits, called α, β, and γ. Following the interaction of the G pro-tein with the ligand–receptor complex, a series of conformational changes are derived, leading to the exchange of GDP by GTP and to the dissociation of the γ-GTP complex.

Such a complex then interacts with an intracellular effector, such as adenylate cy-clase, resulting in its activation with the consequent formation of the secondary messen-ger cAMP (Fig. 5.5) or inositol 1,4,5-triphosphate (IP$_3$), respectively. Myoinositol or inositol

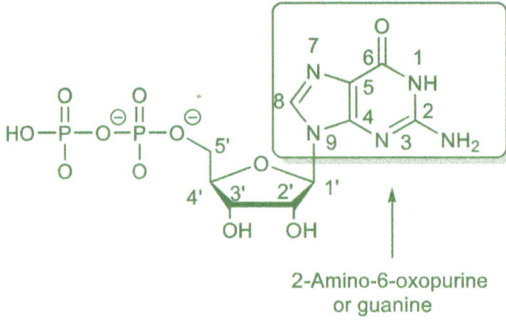

Fig. 5.4: Guanosine diphosphate (GDP).

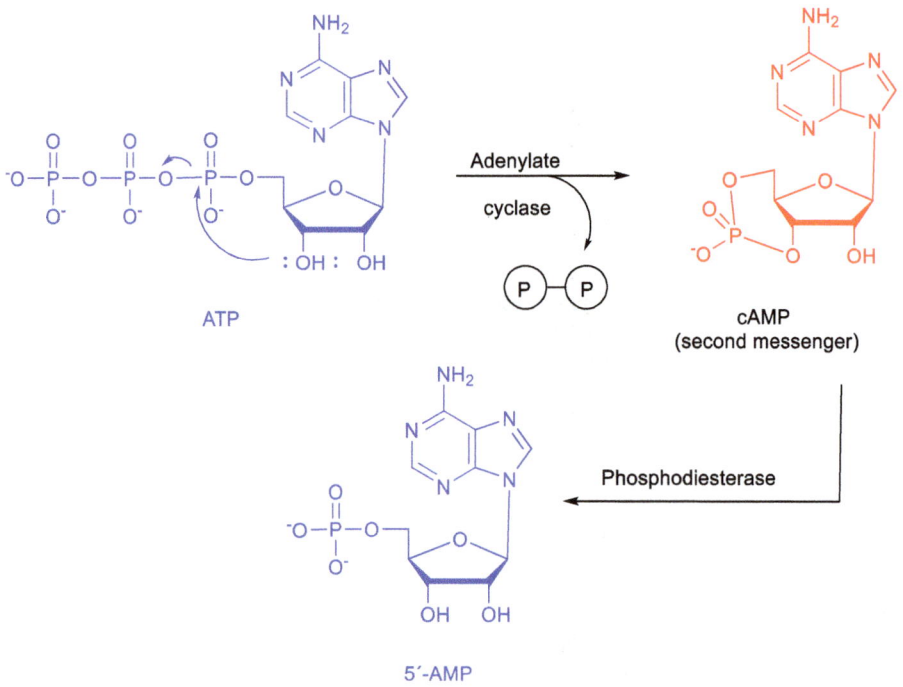

Fig. 5.5: cAMP structure, degradation, and biosynthesis.

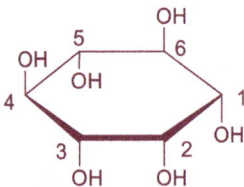

Fig. 5.6: Myoinositol or inositol.

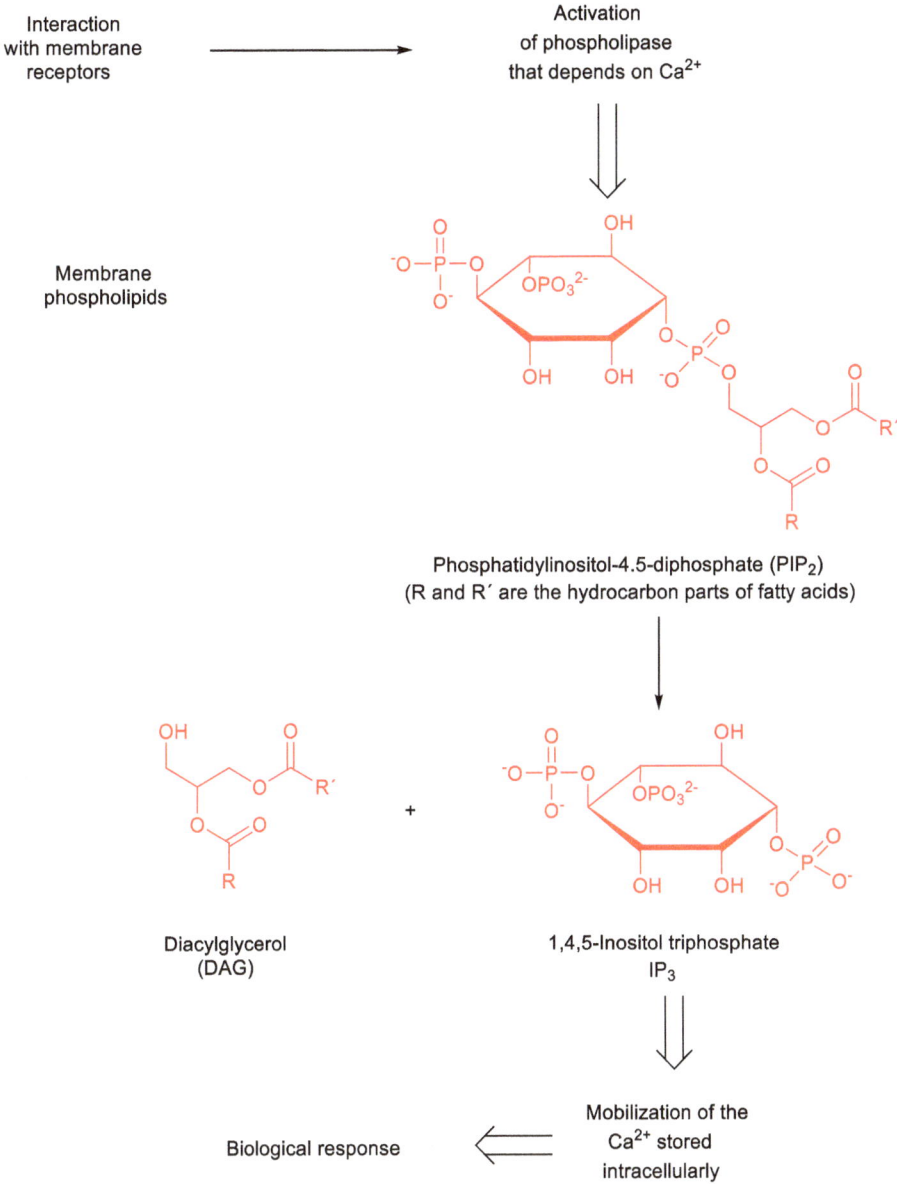

Fig. 5.7: System formed by phospholipids of inositol.

structures are shown in Fig. 5.6. The cAMP is responsible for the activation of protein kinases, which in turn activates certain enzymes by phosphorylation. IP_3 results in the emptying of the Ca^{2+} storage vesicles, whose increase in the intracellular level is responsible for effects as diverse as smooth muscle contraction, glandular secretion, or the release of certain neurotransmitters (Fig. 5.7). Finally, the hydrolysis of a phosphate unit of the α-GTP complex leads to the regeneration of the α-GDP subunit and its combination with the β- and γ-subunits to give rise to the functional G protein again (Fig. 5.8).

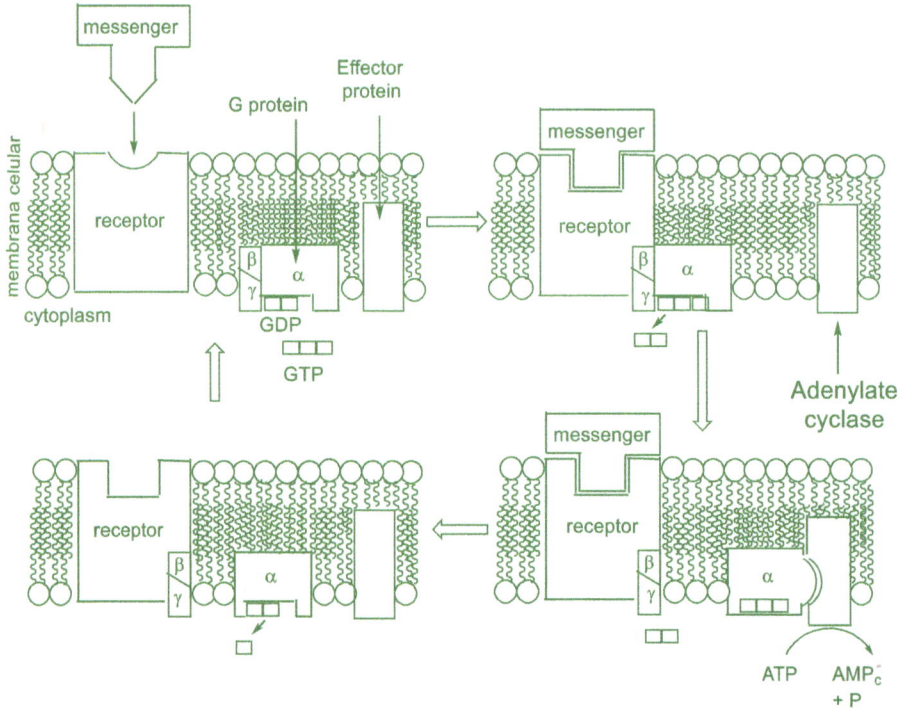

Fig. 5.8: Formation of a secondary messenger by activation of a G protein.

Examples: muscarinic acetylcholine receptor, dopaminergic receptors, and opioid receptors.

5.4.4 Intracellular receptors

These are characteristics of steroid hormones (Fig. 5.9):

Intracellular receptors are specific to steroid hormones (sex hormones, glucocorticoids, vitamin D), and thyroid hormones. Steroid receptors are located at the level of the cell nucleus in certain chromatin sequences leading to the initiation of transcription and protein synthesis.

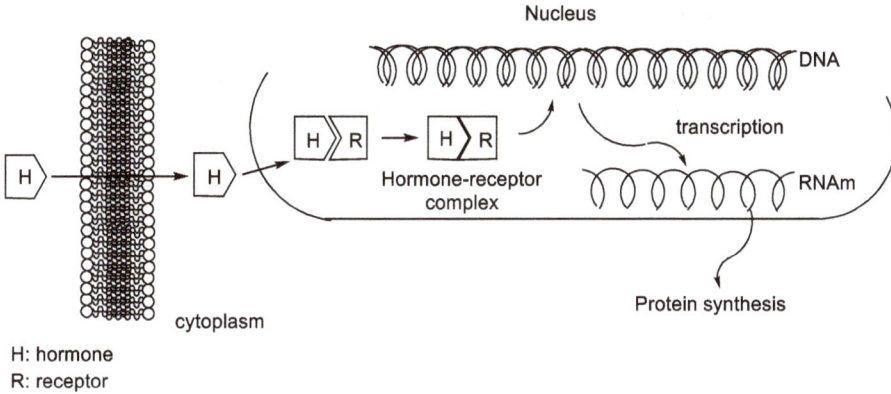

Fig. 5.9: From the messenger to the control of gene transcription.

5.5 Types of bonds: energy range (kcal/mol) per interaction

5.5.1 Intramolecular bonding interactions

– **Ionic bond**: Weak (−5 kcal/mol) reversible. It takes place between charged groups of the drug and the receptor (e.g. ammonium groups, carboxylate, amino, and sulfonamido). In the reinforced ionic bond, hydrogen bonds can also be established between the charged groups (Fig. 5.10). It may seem strange to consider the ionic bond as a weak bond, when a prototype ionic salt is NaCl, which has a melting point of 801 °C. The reason is based on the following reasons: The strong electrostatic force of attraction in ionic compounds reaches out in all directions, and each ion in a solid ionic, crystalline substance is surrounded by other ions of opposite charge. For example, in a sodium chloride crystal, each Na^+ ion is surrounded by six Na^+ ions. Any changes that require disrupting the arrangement of ions in a crystalline ionic compound require a large amount of energy.

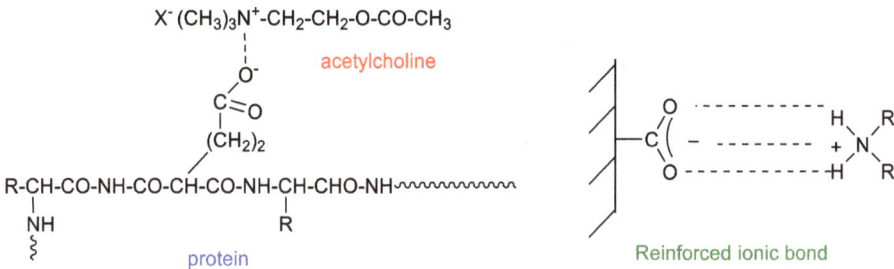

Fig. 5.10: Ionic bonds.

– **Covalent bond**: Strong (–40 to –110 kcal/mol), irreversible). It takes place when the receptor is inactivated by an irreversible antagonist (Fig. 5.11). The covalent bond constitutes the bond of greatest strength in the scale of energies of interaction between two molecules. Due to its stability, this link can be considered practically irreversible.

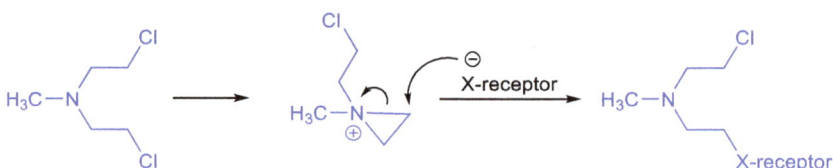

Fig. 5.11: Formation of an irreversible covalent bond.

5.5.2 Intermolecular interactions

– **Hydrogen bond**: Weak (–1 to –7 kcal/mol), reversible. It is established between a weakly acidic hydrogen atom, bonded to an electronegative atom by a covalent bond, and a base that acts as an electron donor (Fig. 5.12). It takes place between the hydrogen of –OH (alcohols and phenols), –NH (amines), –SH (thiols), and a group having not shared pair of electrons (amine, alcohol, ether, and carbonyl).

Fig. 5.12: Hydrogen bond.

– **Dipole–dipole interaction**: Weak (–1 to –7 kcal/mol), reversible. It is polarized by the different electronegativity between the atoms involved (Fig. 5.13).

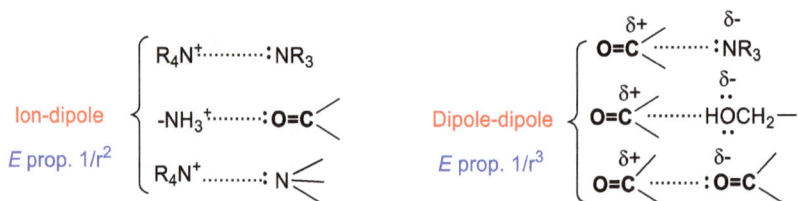

Fig. 5.13: Dipole–dipole interactions.

– **Van der Waals bond**: Weak (–1 to –5 kcal/mol). The set of forces that are established between molecules or atoms, and which cannot be considered as covalent bonds or purely ionic bonds are collectively referred to as "Van der Waals forces" (E prop. $1/r^6$). It is a very weak bond, reversible.

(a) **London dispersion forces**: The London dispersion force is a temporary attractive force that results when the electrons in two adjacent atoms occupy positions that make the atoms form temporary dipoles. In other words, even molecules with no permanent dipole moment have, due to the movement of their electrons, local dipole moments which induce dipoles in the opposite molecule, leading to fluctuating electrostatic attractions. This force is called sometimes a dipole-induced dipole attraction. London forces are the attractive forces that cause nonpolar substances to condense to liquids and to freeze into solids when the temperature is lowered sufficiently (Fig. 5.14). Consider, for example, the case of noble gas atoms, which have a stable electronic configuration:

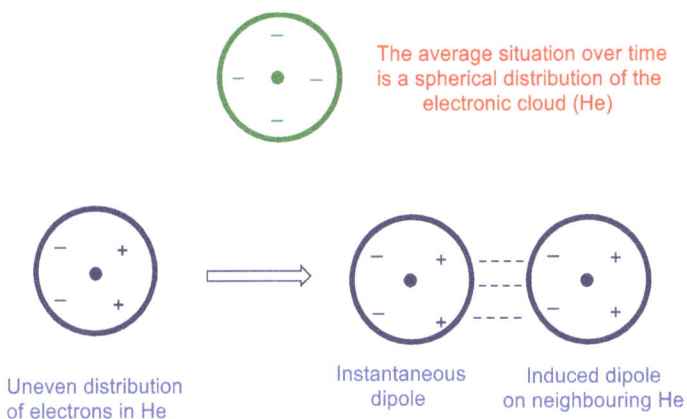

The average situation over time is a spherical distribution of the electronic cloud (He)

Uneven distribution of electrons in He

Instantaneous dipole

Induced dipole on neighbouring He

Fig. 5.14: London dispersion forces.

(b) **Keesom orientation forces**: They exist in molecules of type AB, where A and B do not have the same electronegativity. They exist in molecules with a permanent dipole (Fig. 5.15).

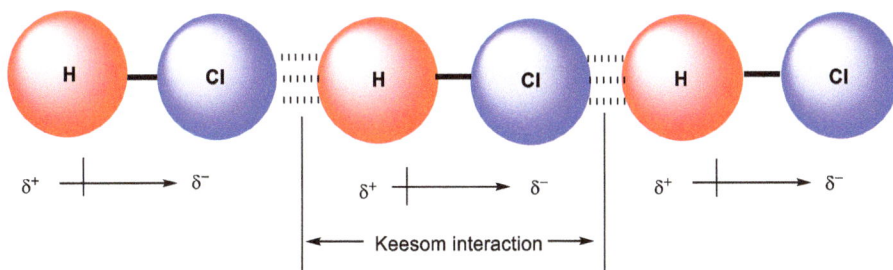

Keesom interaction

Fig. 5.15: Keesom orientation forces.

(c) **Debye induction forces**: They include the force between a permanent dipole and a corresponding induced dipole: CO and Ne.

– **Hydrophobic bond** (−1 kcal/CH_2 group). In the aqueous media, the Van der Waals bond is reinforced by the entropic variation of the system resulting from the desolvation that occurs when two organic molecules approach by their lipophilic part, with the consequent increase of the entropy of the system (Fig. 5.16). Although the hydrocarbon moieties are ordered, a considerably larger number of molecules become disordered, with the consequent increase of the positive value of ΔS and negative value of ΔG. This reinforcement of the Van der Waals bond constitutes the so-called hydrophobic bond of great importance in biological media.

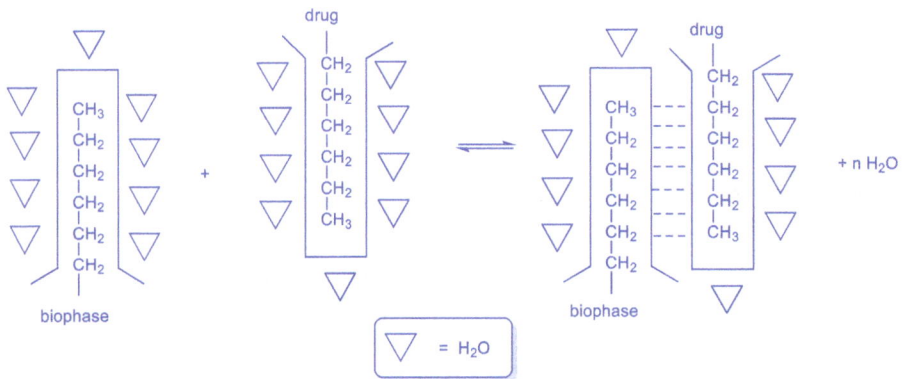

$\Delta G = \Delta H - T\Delta S$

For a process to be spontaneous, $\Delta G < 0$, that is to say, $\Delta H < 0$ ó $\Delta S > 0$

Fig. 5.16: Entropic increase derived from the desolvation required for binding of a drug to its biological target.

– **Charge-transfer complex**. A very weak bond (−1 to −7 kcal/mol). The charge-transfer complexes are formed by the electrostatic attraction between an electron-releasing molecule and an acceptor molecule; although the ionic form that can be represented as donor$^+$–acceptor$^-$ contributes only slightly to the resonance hybrid of these complexes, it increases their stability (Fig. 5.17).

These complexes can be considered as an ionic pair, with the property of undergoing an observable transition of charge between the donor that acquires positive charge and the acceptor that acquires a negative one.

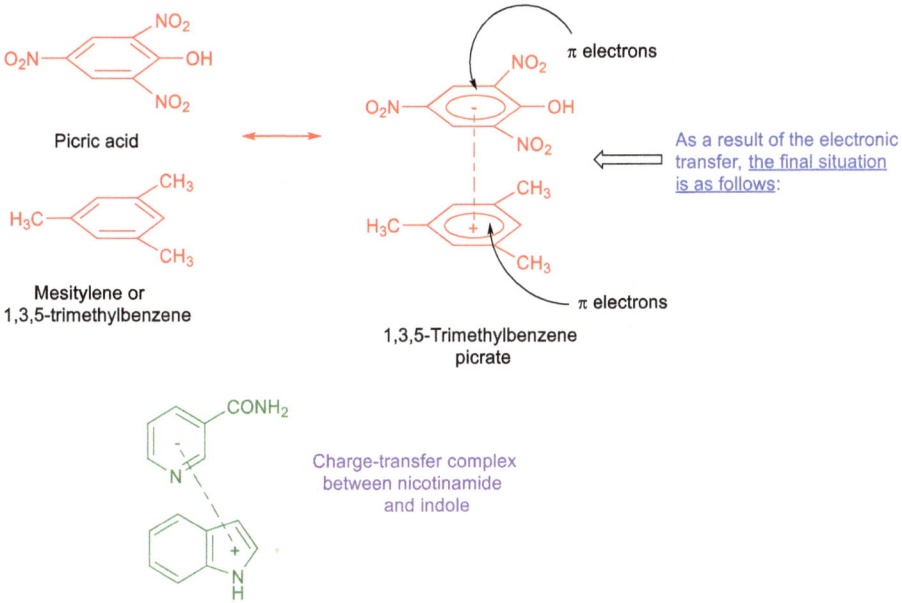

Fig. 5.17: Examples of charge-transfer complexes.

Hydrogen bond is a particular case of charge transfer occurring through a weakly acidic hydrogen atom, attached to the acceptor molecule in a covalent form, and a base that acts as a donor of the electrons that will establish the hydrogen bond.

In the form of a summary and emphasizing the most important interactions that we are going to find from this moment on, Fig. 5.18 represents the type of the most frequent interactions of procaine with the biophase acceptor zones.

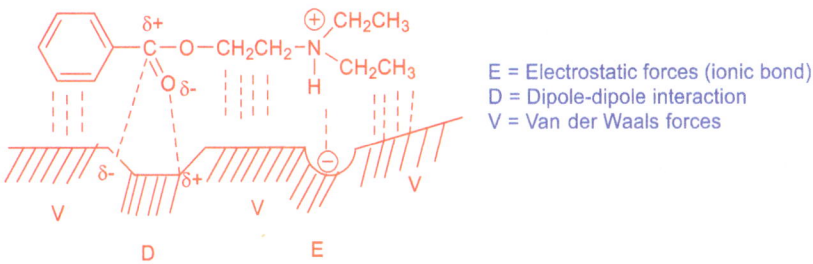

Fig. 5.18: Types of chemical interactions between procaine and the active site of its receptor.

5.6 Conformation and activity: use of rigid analogs

5.6.1 Nomenclature of conformations (Newman projections)

Generally, *n*-butane has four conformers, which are represented in Fig. 5.19 in the Newman projection. Those that rapidly interconvert at room temperature; they cannot be separated. They result from rotation about C–C single bonds. Rotation around a C–C sigma bond is not completely free. There is a possibility of weak repulsive interactions between the bonds on adjacent carbon atoms. Such type of repulsive interaction is known as torsional strain. There are two types of nomenclatures with respect to the different conformations of *n*-butane: the first and older one, located below each of them, and the second, more modern and increasingly used.

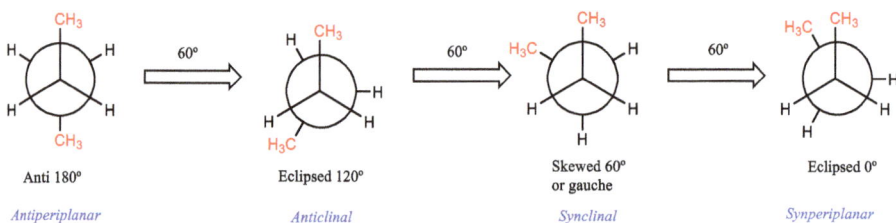

| Anti 180° | Eclipsed 120° | Skewed 60° or gauche | Eclipsed 0° |
| *Antiperiplanar* | *Anticlinal* | *Synclinal* | *Synperiplanar* |

Fig. 5.19: Four conformations of *n*-butane.

- The *anti*-prefix is used when the bonds of the most voluminous groups (in this case, the two methyl groups) form angles higher than 90°.
- The *syn*-prefix is used when the bonds of the most voluminous groups (in this case, the two methyl groups) form angles lower than 90°.
- The *periplanar* termination is applied when the two largest groups are in the same plane.
- The *clinal* termination is when the two most voluminous groups are in different planes.

Conformation is defined as the nonidentical spatial ordering of the atoms of a molecule due to its rotation around one or more single bonds. Let us assume that situation (a) represents the active conformation of an agonist that binds to the receptor through its A and B groups, which causes the biological response through its C group, after interacting with the corresponding complementary groups A', B', and C' of the receptor. Situation (b) represents an antagonist molecule, since it is capable of binding to the receptor, but because it does not have the C group, it would be incapable of giving any biological response. Finally, (c) would represent an isomer of the molecule of (a), with antagonistic activity, for not having the group C in the proper arrangement to interact with C' of the receptor (Fig. 5.20).

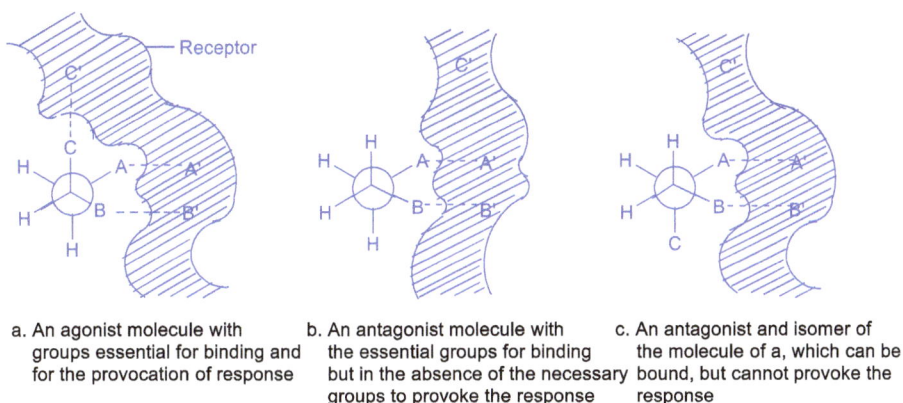

a. An agonist molecule with groups essential for binding and for the provocation of response

b. An antagonist molecule with the essential groups for binding but in the absence of the necessary groups to provoke the response

c. An antagonist and isomer of the molecule of a, which can be bound, but cannot provoke the response

Fig. 5.20: Active conformation of an agonist (a), and conformations of two antagonists (b) and (c) interacting with a hypothetical receptor.

The pharmacophore conformation is not necessarily the preferred conformation in crystalline or dissolution state and may be a thermodynamically unstable conformation. In some cases, the energy of binding to a receptor can compensate for the barrier of forming an unstable conformation (Fig. 5.21).

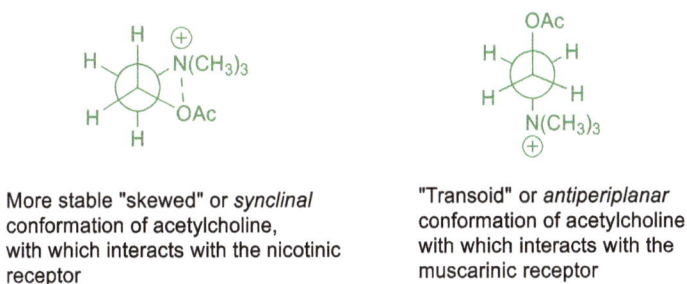

More stable "skewed" or *synclinal* conformation of acetylcholine, with which interacts with the nicotinic receptor

"Transoid" or *antiperiplanar* conformation of acetylcholine with which interacts with the muscarinic receptor

Fig. 5.21: Preferred "skewed" or *synclinal* conformation of acetylcholine, and the transoid or *antiperiplanar* one that acetylcholine acquires upon acting on the muscarinic receptor.

One of the most commonly used methods for the determination of the active conformation of flexible drugs consists in the study of rigid analogs, in which the conformational possibilities are partially restricted.

Decalin has 10 carbons in total. However, remember that any time we have two substituents on a cyclohexane ring (as we do here), it is essential to draw in the stereochemistry in order to avoid ambiguity. Two stereoisomers are possible here: one where the hydrogens at both ring junctions are *cis*, and the other where they are *trans*. The complexity comes when we examine their most stable three-dimensional structures. Each six-membered ring will adopt a chair conformation. The *cis*- and *trans*-stereoisomers of decalin have remarkably different shapes (Fig. 5.22).

cis-Decalin

trans-Decalin

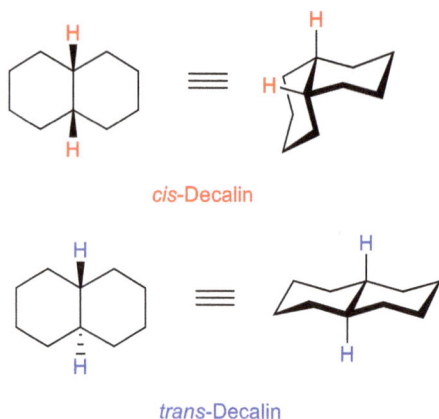

Fig. 5.22: *Cis-* and *trans*-decalins.

The formation of cycles is one of the most frequently used methods for the study of the active conformation of flexible molecules. A classic example of cycle formation is the study carried out on various acetylcholine analogs, in which various conformations are mimicked (Fig. 5.23).

While analogs derived from decalin were devoid of cholinergic activity, cyclopropane analogs support the theory that the *antiperiplanar* conformation is involved in the interaction with the muscarinic receptor and in addition, it is more easily hydrolyzed by acetylcholinesterase.

The lack of activity of the decalin derivatives reveals one of the limitations of the use of rigid analogs. Thus, additional atoms and bonds that are introduced to provide rigidity to the structure can give rise to important changes in the physical and chemical properties with respect to the original molecule.

A thorough study of the antianxiety drug 4-(4-hydroxypiperidino)-4'-fluorobutyrophenone has been carried out using conformers in order to determine the possible steric requirements of the hydroxyl group (Fig. 5.24).

We will only represent the chair with the OH in equatorial or axial position (Fig. 5.25).

Boat and twist-boat conformations have been excluded because of their high energy.

When subjected to muscle relaxation tests, they presented the following order of activity or relative power **b** > **c** > **a**, which implies that the oxygenated function should preferably be in the axial position: the probable pharmacophore conformation would be **b**, i.e. the conformer with the axial hydroxyl group (Fig. 5.26).

The structure of a prototype can be manipulated by limiting its conformational freedom around certain single bonds, as can be seen in the hypotensor clonidine. It is a α_2-adrenergic agonist. Its structure is characterized by having an imidazolidine subunit attached to an aza-styrene moiety with two chlorine atoms at *ortho*-positions. Clonidine has a restricted rotation by steric hindrance and is 30 times more active

This is the closer relationship between the two spatial layouts

It has 8 more carbon atoms   Only one more carbon atom

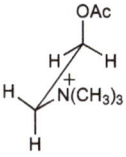

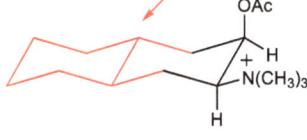

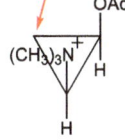

a. "Skewed" or *synclinal* conformation of acetylcholine

b. *trans*-Decaline: rigid analog of the "skewed" or *synclinal* conformation of acetylcholine

c. Rigid cyclopropyl analog of the eclipsed conformation of acetylcholine

Sawhorse projections

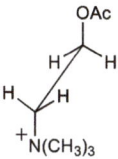

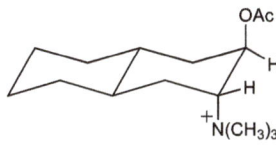

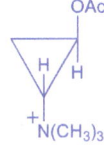

d. "Transoid" o *antiperiplanar* conformation of acetylcholine

e. *trans*-Decaline: rigid analog of the "transoid"or *antiperiplanar* conformation acetylcholine

f. Rigid cyclopropyl analog of the "transoid" or *antiperiplanar* conformation of acetylcholine

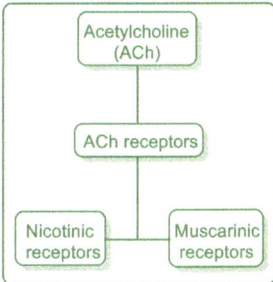

⇩

The *antiperiplanar* conformation is involved in the interaction with the muscarinic receptor

Acetylcholine (ACh)

ACh receptors

Nicotinic receptors

Muscarinic receptors

Fig. 5.23: Acetylcholine cyclic analogs.

Fig. 5.24: 4-(4-Hydroxypiperidino)-4'-fluorobutyrophenone.

than its isomer with free rotation around the nitrogen–phenyl bond (with the two chlorine atoms at positions 3 and 4). These facts can be interpreted, taking into account that the conformational constraint fixes the optimal geometry and diminishes the possible interactions with other targets. Clonidine is a very basic substance (pK_a =

a. Chair conformation with
   the equatorial -OH group

b. Chair conformation with
   the axial -OH group

$$R = -(CH_2)_3CO-\bigcirc-F$$

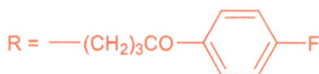

Fig. 5.25: Chair conformations of 4-(4-hydroxypiperidino)-4'-fluorobutyrophenone.

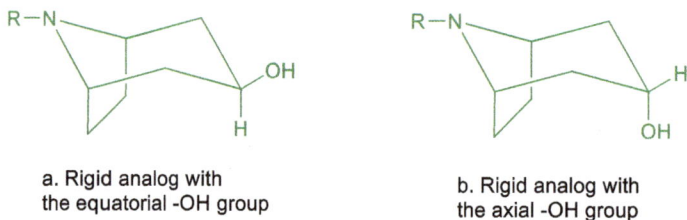

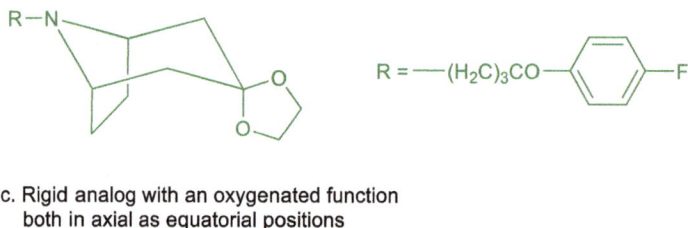

a. Rigid analog with
   the equatorial -OH group

b. Rigid analog with
   the axial -OH group

$$R = -(H_2C)_3CO-\bigcirc-F$$

c. Rigid analog with an oxygenated function
   both in axial as equatorial positions

Fig. 5.26: Use of conformers in order to determine the possible steric requirements of the hydroxyl group.

13); therefore, its protonated form is predominant in the biophase. X-ray studies confirm the exo-imine of the cyclic guanidine moiety of clonidine, in which the *ortho*-dichloro aromatic ring lies on a plane different from that of the imidazoline system with an angle of 75° (Fig. 5.27).

Molecular modeling studies indicate that there is a distance of 0.51 nm between the two nitrogen atoms of the imidazolidine system and the center of the aromatic ring, similar to the distance that occurs in adrenaline (a natural agonist of the α-adrenergic receptors), between the NH_3^+ group and the center of the catechol ring (Fig. 5.28).

Clonidine   Isomer of clonidine

Predominant conformation
of clonidine

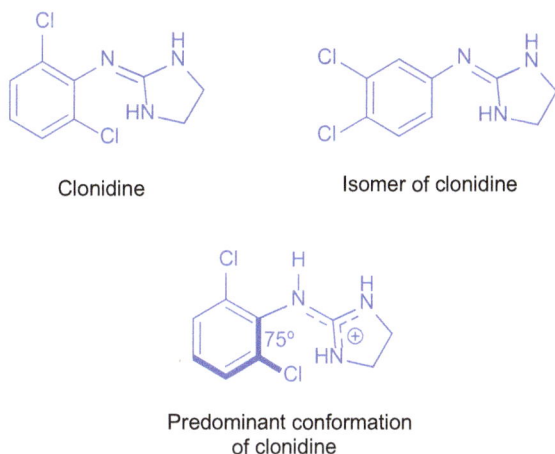

Fig. 5.27: Clonidine (predominant conformation according to X-ray studies) and its less active isomer.

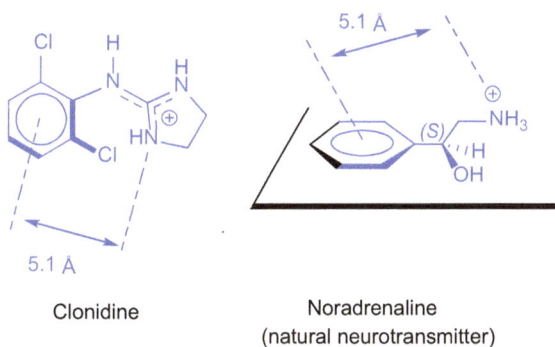

Clonidine   Noradrenaline
(natural neurotransmitter)

Fig. 5.28: Similar distances occur between the nitrogen atoms of the five-membered guanidino moiety and the center of the benzene ring in clonidine, and the ammonium group and the center of the catechol moiety of noradrenaline.

5.7 Absolute configuration and activity: difference between enantiomers

It has been implicitly established in the previous section that the three-dimensional orientation of the functional groups is fundamental for a correct adaptation to the receptor. For example, the bronchodilator activity of (R)-isoprenaline is about 800 times greater than that of its (S)-enantiomer. Similar stereoselectivity is also observed between the enantiomers of catecholamines noradrenaline and adrenaline (Tab. 5.1).

A simple explanation of the different activity of the enantiomers is provided by the Easson–Stedman hypothesis according to which, if one considers that in the adap-

Tab. 5.1: Differences in the bronchodilator activity among catecholamines.

	R = H Noradrenaline
	R = Me Adrenaline
	R = Pr^i Isoprenaline

	Relative bronchodilatation to the (R)-(−)-noradrenaline
(R)-(−)-Noradrenaline	100
(S)-(+)-Noradrenaline	1.4

RIS = 71

| (R)-(−)-Adrenaline | 5,800 |
| (S)-(+)-Adrenaline | 130 } |

RIS = 45

| (R)-(−)-Isoprenaline | 27,000 |
| (S)-(+)-Isoprenaline | 33 } |

RIS = 818

tation of the stereogenic center to the receptor requires at least the interaction through three points, only one of the enantiomers may simultaneously establish such interactions (Fig. 5.29).

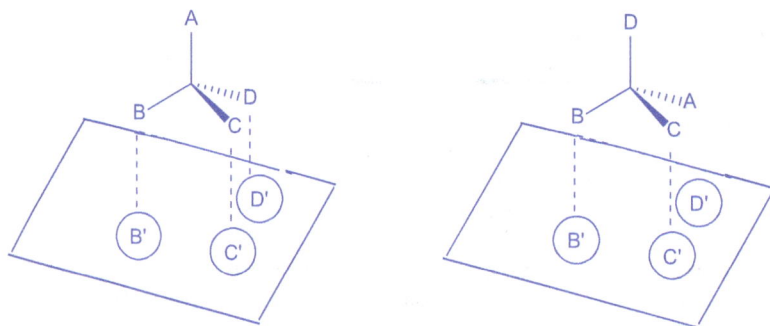

Fig. 5.29: Three-point interaction model.

The more active enantiomer is referred to as eutomer and the least as distomer; the relation of activities between both is called eudysmic ratio.

5.8 Relative configuration and activity

Similarly, alterations in the relative configurations of the substituents of an aliphatic cyclic derivative or an olefin may have repercussions for the recognition by the receptor, as it could lead to a loss of complementarity and accordingly reduction of affinity and intrinsic activity (Fig. 5.30).

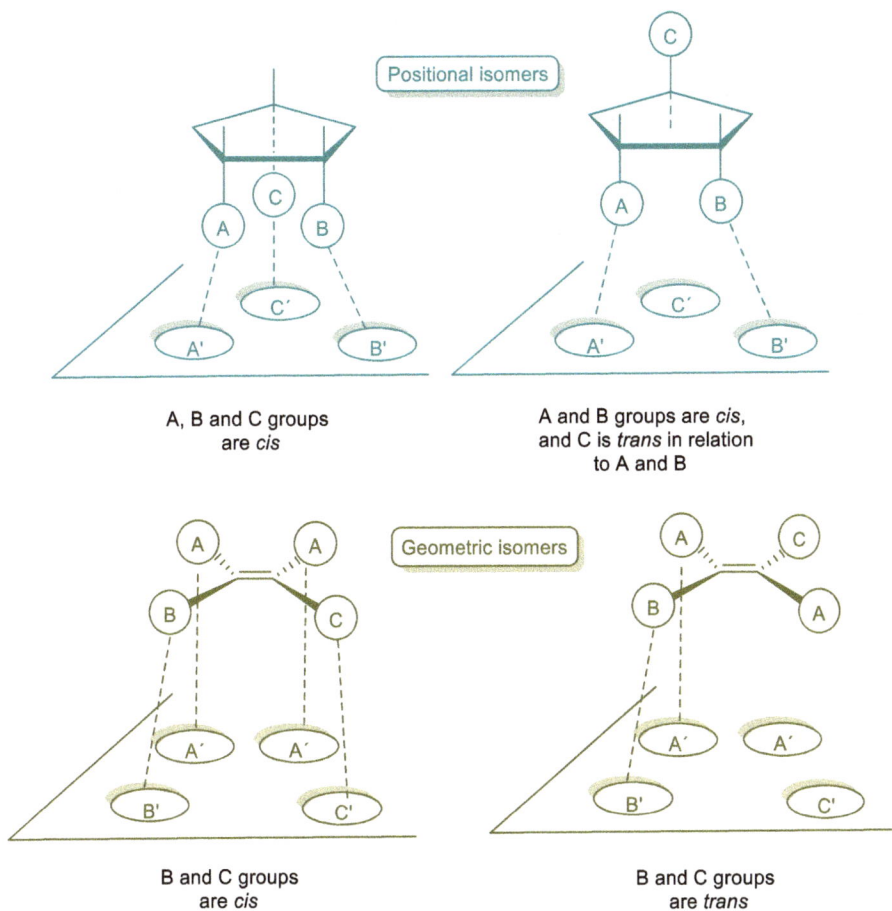

Fig. 5.30: Relative configurations and ligand-receptor molecular recognition.

As an example, we have the case of cisplatin, which is an important anticancer agent, while its *trans*-isomer (transplatin) does not present any useful pharmacological activity (Fig. 5.31).

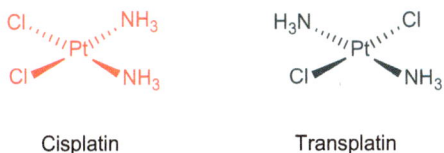

Cisplatin   Transplatin

Fig. 5.31: Cisplatin and transplatin.

Sulindac is a nonsteroidal anti-inflammatory drug (NSAID) of the arylacetic acid class that evidences the importance of the configuration of the double bond in the biological activity. The (Z)-isomer is the active drug, while the (E)-isomer or isosulindac is less active.

This difference in activity between the two geometric isomers led to the identification of the bioactive conformation of indomethacin, another NSAID of the group of aryl acetic acids with an indole nucleus. Indomethacin was the prototype for the development of sulindac. However, in presenting an amide function involving a p-chlorobenzoyl residue attached to the indole nitrogen, it is possible that it undergoes enzymatic hydrolysis, which gives rise to certain Central Nervous System (CNS) side effects. Therefore, sulindac was designed, and a new bioisosteric relationship was discovered between the indomethacin indole system and the indene ring of sulindac. When considering the structure of the active diastereoisomer of sulindac [(Z)-isomer)], it was suggested that the **A** conformation is the bioactive conformation of indomethacin (Fig. 5.32).

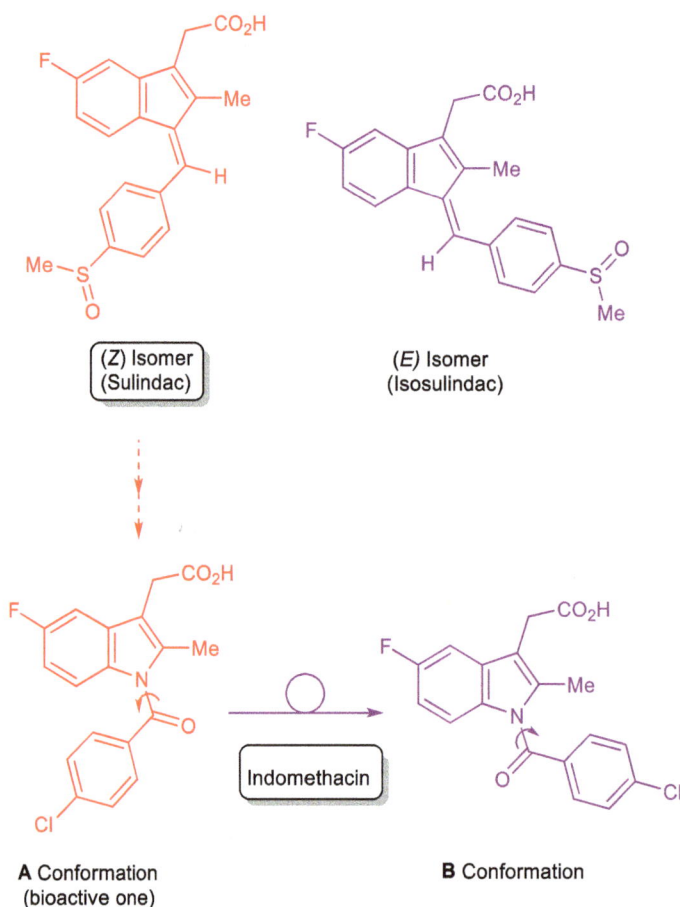

Fig. 5.32: Sulindac, isosulindac, and bioactive conformation of indomethacin.

5.9 Key notes: receptors

Receptors: Most receptors are proteins that traverse the cell membrane with a binding site on the extracellular region. Binding of a chemical messenger causes the receptor to change shape, initiating a process that results in a message being received by the cell. The messenger does not undergo any reaction and departs unchanged, allowing the receptor to reform its original shape.

Chemical messengers: Chemical messengers are neurotransmitters or hormones. Neurotransmitters are released by nerves to interact with specific target cells. Hormones are released by glands and travel around the body to interact with all the receptors that recognize them.

Binding site: The binding site of a receptor is the equivalent of an enzyme's active site, but has no catalytic activity.

5.10 Key notes: binding interactions

Agonists and antagonists: Agonists mimic a receptor's chemical messenger. Antagonists bind to a receptor but do not activate it. By binding to the receptor, they prevent activation by the natural messenger.

Dipole–dipole interactions: Dipole–dipole moments may be important in orientating a molecule when it enters a binding site. The dipole moment of the drug may align itself with localized dipole moments present in the binding site. If the alignment is such that the binding groups are correctly positioned, then the drug is more likely to bind and will have a good activity.

Covalent bonds: Some drugs form covalent bonds to their targets. Alkylating agents react with nucleophilic groups such as serine, cysteine, and guanine, leading to the formation of a covalent bond and the irreversible inhibition of the target.

5.11 Key notes: stereochemistry

Introduction: Stereochemistry plays an important part in pharmaceutical chemistry. Drugs must be the correct shape to fit binding sites. They must also have their binding groups in the correct relative positions to interact with groups in the binding site.

Isomers: Constitutional isomers are different structures that have the same molecular formula. Configurational (or geometrical) isomers have the same atoms and bonds, but have different shapes, which cannot be interconverted through simple bond rotation. Optical isomers are configurational isomers that can exist as two nonsuperimposable mirror images. Conformational isomers are different shapes of the same compound, interconvertible by single bond rotation. The active conformation is the shape adopted by a drug when it binds to its binding site.

Chirality and asymmetric centers: Chirality is defined as the asymmetry of a molecule. In order to be chiral, a molecule must have no more than one axis of asymmetry. Asymmetric centers are carbon atoms with four different substituents. The enantiomers of a chiral drug can interact differently with chiral targets such as proteins.

5.12 Exercises

1. Indicate the possible binding forces that will bind the following drugs to their receptor:

(a) Levodopa

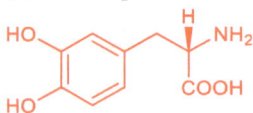

(d) α-Methylnoradrenaline

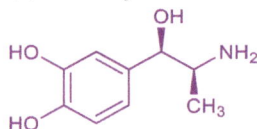

(b) Neostigmine

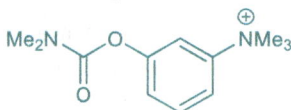

(e) Sumatriptan

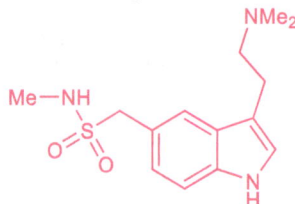

(c) Warfarin

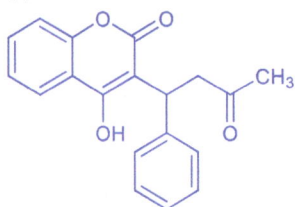

(f) Camazepam

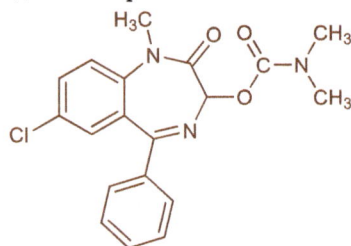

(g) Captopril

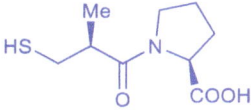

Me

HS

N

O

COOH

(h) Dextromoramide

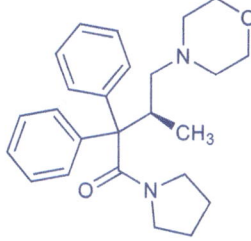

O

N

CH₃

O

N

2. Indicate how a covalent bond will be formed between the following drugs and their specific receptors:

(a) Melphalan

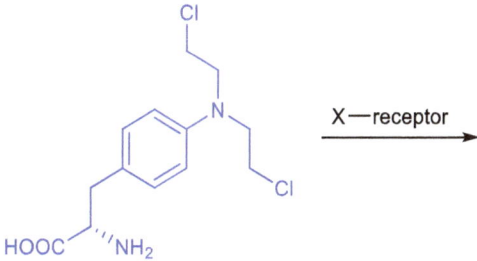

Cl

N

Cl

X—receptor

HOOC ''NH₂

(b) Fluostigmine

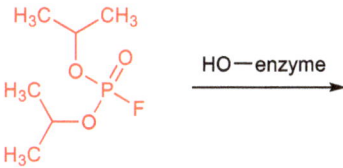

H₃C CH₃

O O

H₃C P F

O

H₃C

H₃C

HO—enzyme

(c) Ethacrynic acid

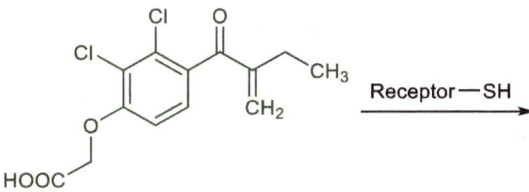

Cl O

Cl

CH₃

O CH₂

HOOC

Receptor—SH

(d) Cefaclor

Protein—OH
$\longrightarrow$

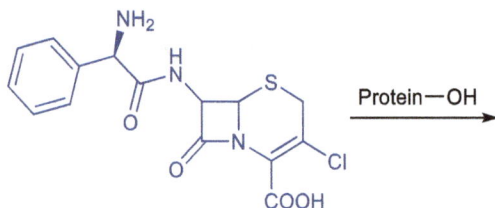

(e) Phenoxybenzamine

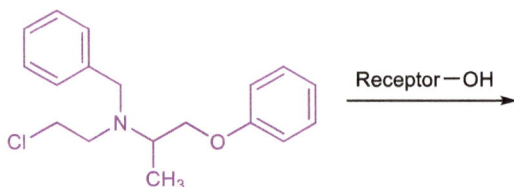

Receptor—OH
$\longrightarrow$

3. Propranolol is a β-adrenergic receptor antagonist that acts as a cardiovascular agent. The *S*-enantiomer is the active isomer, although it is used clinically as the racemate. Draw the structure of this enantiomer and compare it with that of the neurotransmitter (*R*)-noradrenaline, explaining the antagonism. Represent the binding forces that will bind each of these molecules to the β-adrenergic receptor:

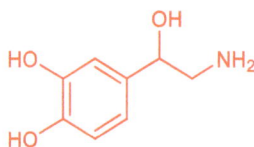

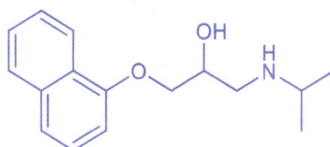

Noradrenaline   Propranolol

4. Propantheline is an acetylcholine antagonist drug. (a) Show the drug–receptor (D-R) interaction zones through the different binding forces and compare them with those of acetylcholine. (b) Given the following interpretation for the D-R interaction responsible for the pharmacological response, compare the values of the rate constants for an agonist such as acetylcholine, an antagonist such as propantheline, and for an inactive compound:

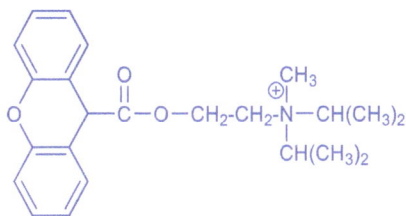

Propantheline

$$D+R \underset{K_2}{\overset{K_1}{\rightleftharpoons}} \quad [DR] \quad \overset{K_3}{\rightarrow} \quad \text{Biological response}$$

5. Justify the possible difference in activity of the following pairs of structures in their binding to the receptor, indicating whether they are conformational isomers, enantiomers, or diastereoisomers:

(a)

Cyclopropane analogs of acetylcholine

(b)

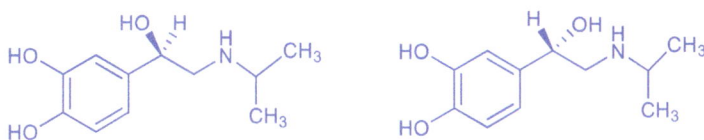

Isoprenaline

(c)

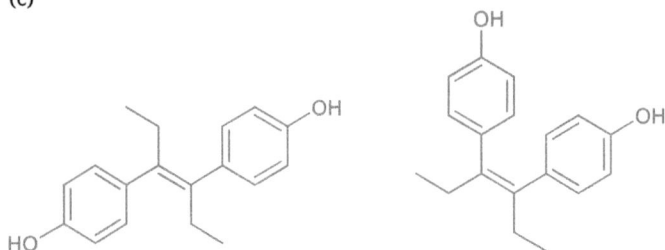

Diethylstilbestrol

(d)

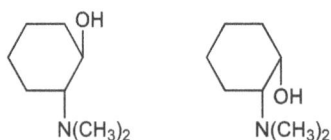

2-Dimethylaminocyclohexanol

(e)

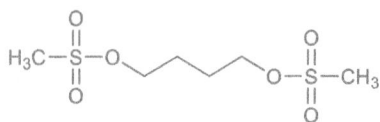

Busulfan (antineoplastic agent)

Rigid analogs of busulfan

(f)

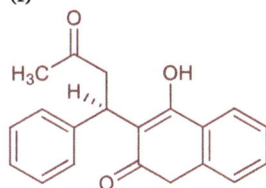

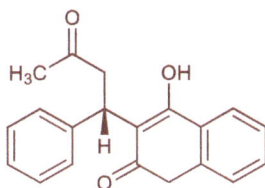

(*R*)-Warfarin

(*S*)-Warfarin

6. Tamoxifen acts as an estrogen receptor antagonist. Suggest how you can join the receptor to manifest that effect. The tamoxifen metabolite, however, acts as an agonist rather than as an antagonist. Why?

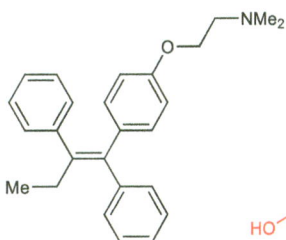

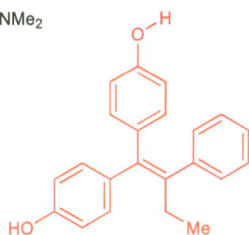

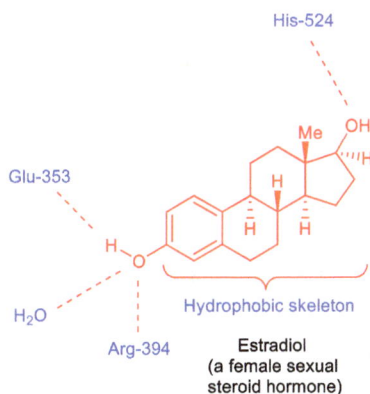

Tamoxifen

Tamoxifen metabolite

Estradiol
(a female sexual
steroid hormone)

6 Quantitative drug design: parameters and quantitative structure–activity relationships

6.1 Goals

- To know the **q**uantitative **s**tructure–**a**ctivity **r**elationship tool as an instrument for optimizing a prototype and reducing its cost
- To know the tool of molecular modeling as a research method for discovering more selective and more economical new drugs
- To know superficially other current tools in the search for new drugs

6.2 Introduction

The discovery of new cancer drugs can happen in different ways:
- Serendipity
- Screening of natural products
- Chemical modification of existing drugs
- Rational drug design or simply rational design

By rational drug design, we mean to obtain the active molecules from a pharmacological point of view, based on previous considerations that predict its activity. The aim of this chapter is to give an introduction to **QSAR** (acronym for **q**uantitative **s**tructure–**a**ctivity **r**elationship) methodology for beginners. I will try to apply QSAR in a proper manner to gain more insight into structure–activity relationships and biological mechanisms.

The historical birth of the **QSAR** idea was in the year 1870, when the biological activity (BA) was postulated to be a function of the chemical structure (6.1):

$$BA = f(\text{chemical structure}) \tag{6.1}$$

BA, measured in the form of log(1/C), where C is the molar concentration of the drug capable of producing a certain activity, is a function of the physicochemical parameters of the molecules:

$$\log(1/C) = f'(\text{electronic + hydrophobic + steric parameters}) \tag{6.2}$$

The year 1964 may be considered as the year of birth of modern QSAR methodology. A widely used method is the Hansch–Fujita method, which establishes a correlation between BA and a linear combination of parameters representing the physicochemical changes within a number of molecules. The dependence of BA with the electronic, hydrophobic, and steric parameters (6.2) is the simplest relation of a great variety of

https://doi.org/10.1515/9783111316901-006

Hansch equations. Equations have been used in the last six decades relating the BA to almost all possible conceivable combinations of lipophilic, polarizability, electronic, steric parameters, with or without other additional indicators (among others, molecular orbital parameters). In this introductory course, we will study only the simplest forms of Hansch equations. Herein, we will study only the simplest forms of Hansch equations.

BA can be measured in isolated enzymes or receptors (in vitro), cells removed from intact animals or humans (ex vivo), or in the whole animal (in vivo). The in vitro assay provides information on the enzyme–drug or receptor–drug interaction and ultimately on the affinity of the drug. Such experiments are very useful in that provide evidence about the biological target. The ex vivo assays provide evidence that the target under study is operational in the species studied, as well as demonstrating drug absorption and penetration to the desired site of action. Research on isolated enzymes, isolated receptors, or isolated cells are the simplest to carry out and provide the clearest insight as to mechanism. However, in the end, only whole animal data can disclose the understanding required in drug design or toxicology research. Compounds with suitable potency, efficacy, and selectivity at the primary target and that have shown activity in disease relevant in in vitro and ex vivo systems also need to have good pharmacokinetics if they are to deliver in vivo efficacy and duration of action.

Most important condition in a QSAR study is that all compounds of a set should have the same molecular frame, i.e. an identical parent compound, with structural variation in only one or several positions.

Figure 6.1 represents linear and second-degree equations:

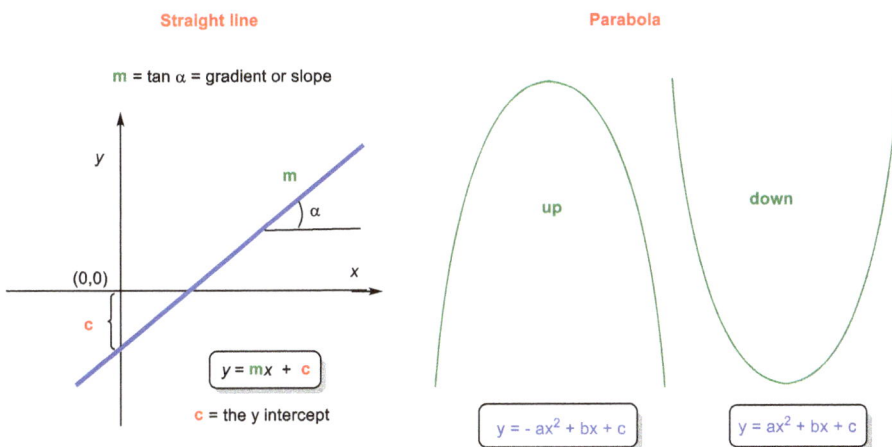

Straight line

$m = \tan \alpha = $ gradient or slope

$y = mx + c$

$c = $ the y intercept

Parabola

up

$y = -ax^2 + bx + c$

down

$y = ax^2 + bx + c$

Fig. 6.1: Linear and second-degree equations.

Since Hammett studies obtained a quantitative expression of the reactivity of organic substances, they represent the background as a function of their structures (Fig. 6.2).

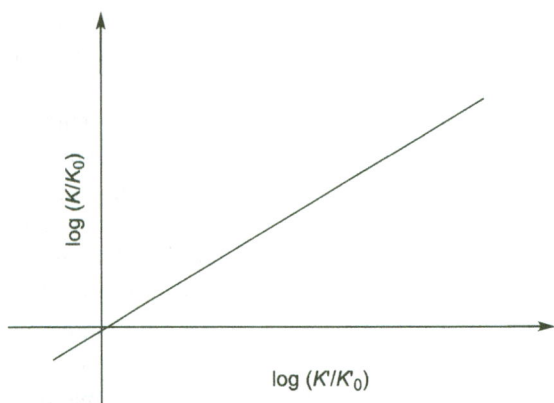

Fig. 6.2: Correlation between the dissociation constants of unsubstituted and substituted benzoic acids (abscissa axis) and those of phenylacetic acids (ordinate axis) with the same substitution patterns (*meta* and *para*).

6.3 Electronic parameters

The abscissa values are calculated from the dissociation constants of unsubstituted and substituted benzoic acids (6.3). On the other hand, the ordinate values are obtained from phenylacetic acids (6.4) with the same substitution patterns (Fig. 6.2). The substitution refers to *meta*- and *para*-positions, and never to *ortho*-positions, because of possible incidence of the steric effect:

$$XC_6H_4CH_2COOH + H_2O \rightarrow XC_6H_4CH_2COO^- + H_3O^+ \tag{6.3}$$

Y-axis values

$$XC_6H_4COOH + H_2O \rightarrow XC_6H_4COO^- + H_3O^+ \tag{6.4}$$

X-axis values.

ρ is the slope of the line. The values of the abscissa axis are always those of benzoic acid and are called σ (Hammett's constant). Therefore, it is possible to write:

$$\log \frac{K}{K_0} = \rho \log \frac{K'}{K'_0} \tag{6.5}$$

K_0 and $K'_{0'}$ represent equilibrium constants of unsubstituted compounds and K or K' of the substituted derivatives (6.5).

Another form is (6.7), in which sigma is defined according (6.6):

$$\sigma = \log \frac{K'}{K'_0} \tag{6.6}$$

$$\log \frac{K}{K_0} = \rho\sigma \tag{6.7}$$

6.3.1 Interpretation of Hammett constants

There are two ways for the interpretation: (a) according to Fig. 6.3 and (b) according to Fig. 6.4.

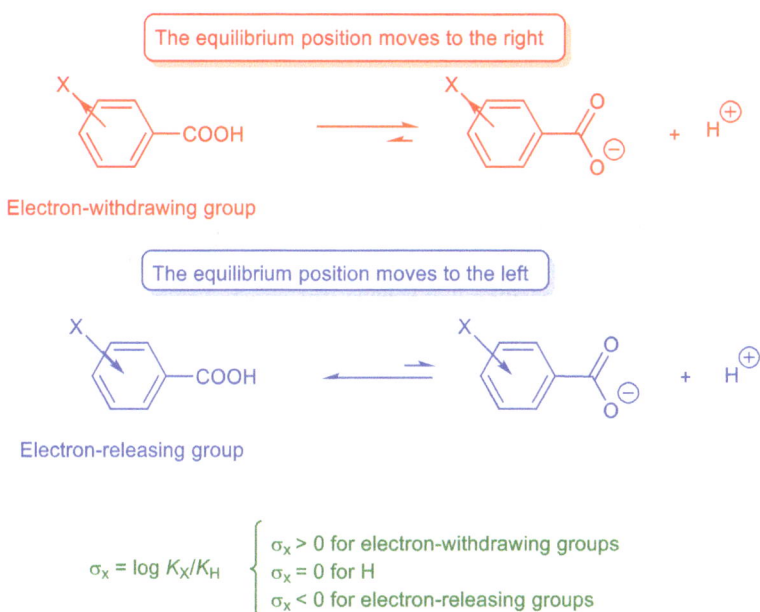

Fig. 6.3: Substituent effects on dissociation of benzoic acids.

When a G group attracts electrons, it stabilizes the carboxylate group and strengthens the acid, while when G releases electrons it exerts the opposite effect (Fig. 6.4).

Therefore, electron-withdrawing substituents have positive σ values, and electron donors have negative values. The hydrogen has a value of $\sigma = 0$.

6.3.1.1 Electronic effects

Trichloroacetic acid is 15,000 times more acidic than acetic acid. The inductive effect of the C–Cl bond is responsible for this difference (Fig. 6.5).

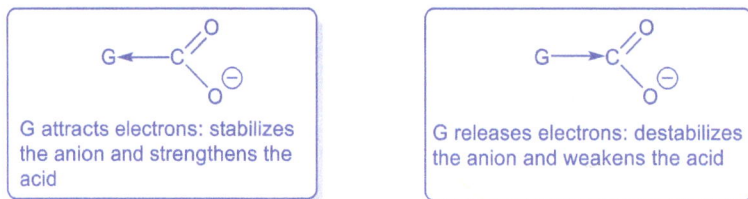

G attracts electrons: stabilizes the anion and strengthens the acid

G releases electrons: destabilizes the anion and weakens the acid

Fig. 6.4: Stabilization or destabilization of the carboxylate anion depending on the electronic nature of G.

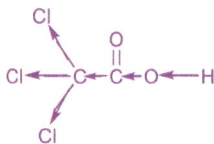

Fig. 6.5: Trichloroacetic acid (TCA) is a strong acid: from a chemical ionization point of view, TCA is a much stronger acid than acetic acid because electronegative chlorine atoms remove electron density from the carboxyl end of the molecule, producing a partial positive charge on the carboxyl moiety.

The inductive effect decreases rapidly with the distance as indicated by the comparison of α-, β-, and γ-chlorobutyric acids with butyric acid. Chlorine in position α increases the acidity nine times. In position β, it reduces the effect of chlorine being the acidity six times higher, and in the position γ, a chlorine is only able to double the strength of the acid (Fig. 6.6).

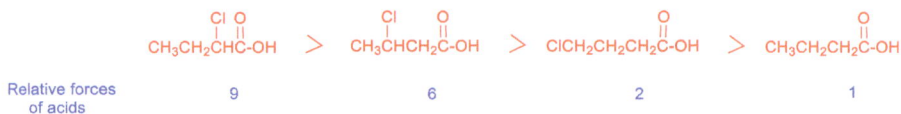

Fig. 6.6: Relative forces of chlorobutyric acids.

The Hammett constant takes into account two electronic effects: the inductive and the resonance or mesomeric effect. Hence, the value of σ of a particular substituent will depend on whether the substituent is *meta* or *para*. This is indicated by the subscript *m* or *p* after σ. For example, the nitro group has a value of $\sigma_m = 0.71$ and $\sigma_p = 0.78$. In the *meta*-position, the attractive electron effect is due to the inductive influence of the substituent, while in the *para*-position, both inductive and resonance effects play a leading role, so the σ_p value is greater (Fig. 6.7).

However, for the hydroxyl group, $\sigma_m = 0.12$ and $\sigma_p = -0.37$. In the *meta*-position it has an electron-withdrawing influence, as a consequence of the inductive effect $-I$. In the *para*-position, the electron-donating influence, due to the resonance of the electron pair of the hydroxyl group, is more important than the electron-withdrawing effect caused by the induction effect $(-I)$ (Fig. 6.8).

A) Nitro group in *meta*: the electronic influence on R is inductive

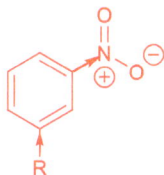

B) Nitro group in *para*: the electronic influence on R is due to resonance and inductive effects

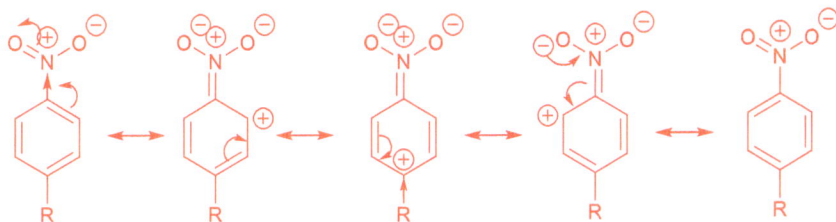

Fig. 6.7: Effects of the nitro group at the *meta*- and *para*-positions.

A) Hydroxyl group in *meta*: the electronic influence on R is inductive

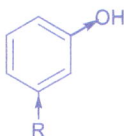

B) Hydroxyl group in *para*: the electronic influence on R is dominated by the resonance effect

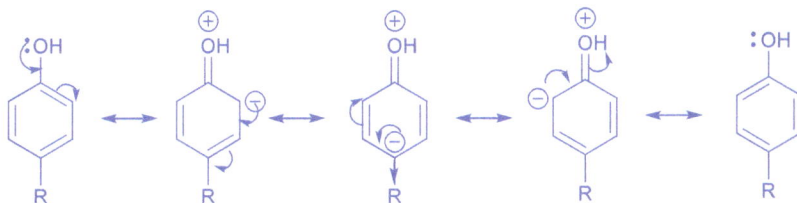

Fig. 6.8: Effects of the phenol group at *meta*- and *para*-positions.

It is said that a group possesses an $-I$ effect if it acquires negative charge by induction; the effect is called $+I$ if the charge acquired by induction is positive; similarly, it is said to have a $-R$ (or $+R$) effect if it acquires negative (or positive) charge by resonance. It is very important not to confuse alkyl group (R) with the resonance effect, which is identified as *R*.

There are limitations as to the electronic constants described so far: for example, the constants of the Hammett substituents cannot be measured for *ortho*-substituents because the possible electronic effect is distorted by the steric effect.

6.4 Substituents classified according to their directing power in electrophilic aromatic reactions

Recognizing substituents as electron-donating (or -releasing) or electron-withdrawing is a useful skill for evaluating reaction mechanisms. For electrophilic aromatic reactions, abbreviated as substitution electrophilic aromatic (S_EAr), although some sources use the acronym EARs, the rate-determining step is the formation of a positively charged sigma complex. Figure 6.9 shows the most common substituents according to their directing power in EAR:

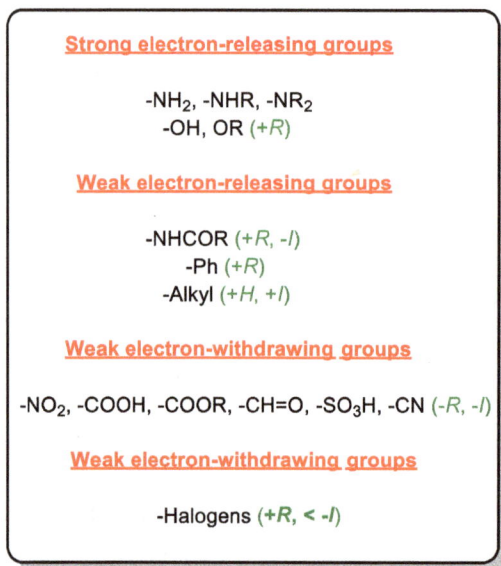

Strong electron-releasing groups

$-NH_2$, $-NHR$, $-NR_2$
$-OH$, OR $(+R)$

Weak electron-releasing groups

$-NHCOR$ $(+R, -I)$
$-Ph$ $(+R)$
$-Alkyl$ $(+H, +I)$

Weak electron-withdrawing groups

$-NO_2$, $-COOH$, $-COOR$, $-CH=O$, $-SO_3H$, $-CN$ $(-R, -I)$

Weak electron-withdrawing groups

$-Halogens$ $(+R, < -I)$

Fig. 6.9: Effect of substituents on electrophilic aromatic reactions.

Overall, groups with a pair of unshared electrons in the atom attached to the benzene ring are activating (in EARs) and *ortho*- and *para*-directing groups ($-NH_2$, $-OCH_3$, etc.). *ortho- and para-directing groups are electron-releasing groups.*

If the atoms attached to the ring support a partial or total positive charge (all resonance structures must be checked), it will be a *meta*-directing group. *meta-directing groups are electron-withdrawing groups.*

The halide ions, which are electron-withdrawing but *ortho*- and *para*-directing groups, are the exception:
–   All activators are *ortho–para*-directing groups.
–   Deactivators (halogens) are *ortho–para*-directing groups.
–   Deactivators (not halogens) are *meta*-directing groups (Fig. 6.10).

Fig. 6.10: *meta*-directing groups.

6.5 Hyperconjugation

Propene is stabilized by the resonance structures in Fig. 6.11.

Fig. 6.11: Resonance structures of propene (*hyperconjugation with a bond loss*).

Such resonance is not very effective, since ionic forms have a covalent bond less than the ordinary forms of Kekulé. These ionic forms will be analogous to forms such as $CH_3^-H^+$ for methane, and obviously, such forms are much less important than those that only have covalent bonds.

Considering that the ionic forms of this type can only be written by "loss" of a covalent bond, this type of resonance is called "hyperconjugation with a bond loss (Fig. 6.11)". However, for an alkyl-substituted carbonium ion, hyperconjugation resonance forms that have the same number of covalent bonds as the main structures can be drawn.

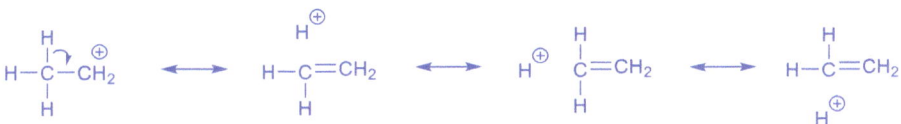

Fig. 6.12: Isovalent hyperconjugation.

This type of resonance is called "isovalent hyperconjugation (Fig. 6.12)" and is much more important than "hyperconjugation with a bond loss".

6.5.1 Electronegativity of an sp^n hybrid orbital

A *p*-orbital has the approximate shape of a pair of lobes on opposite sides of the nucleus, or a somewhat dumbbell shape.

A hybrid orbital sp^n has more s character the lower the value of n. Considering that the s electrons are closer to the nucleus and more fastened than p electrons, the lower the value of n, the more electronegative the orbital.

For instance, an sp^3 orbital consists of 25% s character and 75% p character, and an sp orbital consists of 50% s character and 50% p character. Thus, electronegativity increases in the direction $p < sp^3 < sp^2 < sp < s$.

This influence creates a small dipole that has the negative end next to the double link and the positive next to the alkyl group. *In this way the alkyl groups have an inductive effect giving rise to an electron-releasing effect when bound to an unsaturated carbon.*

6.5.2 Activating effect of the toluene methyl group versus S$_E$Ar reactions

Compared to benzene, toluene is more reactive by virtue of the *inductive and hyperconjugative effects of the methyl group*. Remember that when an sp^3 orbital of a carbon atom is attached to an sp^2 orbital (also of a carbon atom), there is an electronic transfer in the direction of the sp^2 carbon.

Hyperconjugation has little importance in the fundamental state of toluene as it is a "hyperconjugation with a bond loss"; nevertheless, it is much more important in the Wheland intermediate because it becomes an "isovalent hyperconjugation (Fig. 6.13)".

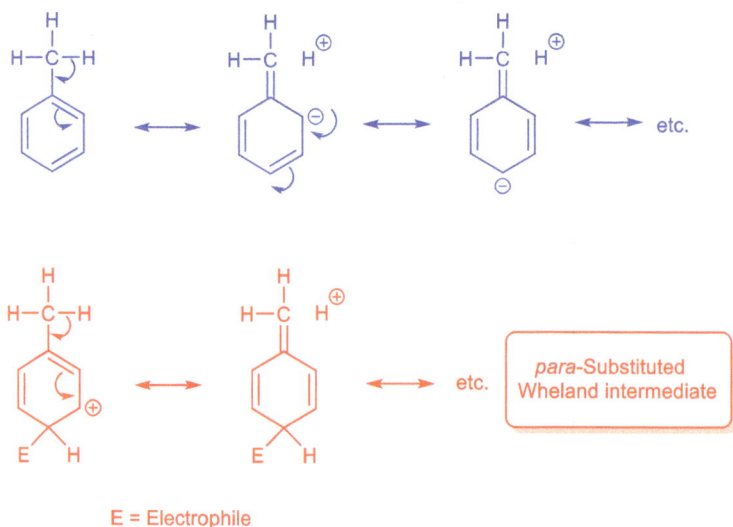

E = Electrophile

The student must represent the corresponding *ortho*-substituted Wheland intermediate

Fig. 6.13: The methyl group releases electrons toward the benzene ring partly due to inductive effect and mainly due to hyperconjugation.

So toluene and alkylbenzenes are *ortho–para*-directing groups and more reactive than benzene versus S_EAr. However, alkyl groups are weak activating groups.

6.5.3 Difference between hyperconjugation and resonance

Hyperconjugation: It is the stabilization effect on a molecule due to the interaction between an s-bond and a π-bond.

Resonance: Resonance is the stabilizing effect of a molecule through delocalization between free electron pairs and π-electrons, or between π-electrons.

Bond lengths:
– **Hyperconjugation:** It causes the σ-bond length to be shortened.
– **Resonance:** It has no effect on the σ-bonds.

To sum up, hyperconjugation *is an extension of resonance since both methods cause the stabilization of a molecule through delocalization of electrons.* However, hyperconjugation involves delocalization of σ-bond electrons along with π-bond electrons, whereas resonance causes the delocalization through interaction between free electron pairs and π-electrons, or between π-electrons.

6.6 Hydrophobic parameters: partition coefficient and hydrophobic substituent constant

Lipophilicity or hydrophobicity of molecules measures the relative tendency they have to prefer a nonaqueous environment against an aqueous one. In drugs, it is decisive for their absorption, distribution, and elimination and also for their bonding with the therapeutic target (hydrophobic bonds). P (6.8) is defined as the ratio of concentrations (C_1/C_2) of a single species between two phases in equilibrium (phase 1 is *n*-octanol and phase 2 is water):

$$P = [\text{drug}]_{\text{octanol}} / [\text{drug}]_{\text{water}} \tag{6.8}$$

However, log P is more useful.

Lipophilicity is an important factor affecting the distribution and fate of drug molecules. Increased lipophilicity has been shown to correlate:
– with increased BA;
– poorer aqueous solubility;
– increased detergency/cell lysis;
– increased storage in tissues;
– more rapid metabolism and elimination: *lipophilic compounds tend to have a greater affinity for metabolic enzymes*;
– increased rate of skin penetration;

- increased plasma protein binding;
- faster rate of onset of action, and in some cases, shorter duration of action.

Drug transport characteristics, or how medications go from the place of application (such as an injection site and the gastrointestinal system) to the site of action, are significantly influenced by log P. Drugs must enter and travel through numerous cells to reach the site of action because they are often transported via the blood. The tissues that a given chemical can reach will therefore depend on the log P. Drugs that are extremely water-soluble will not be able to penetrate lipid barriers and reach lipid-rich organs like the brain and other neuronal tissues. Nevertheless, substances can pass through the "blood–brain barrier" by diffusing from one aqueous phase (blood) to another (cerebrospinal fluid). However, substances that are highly lipophilic will become caught in the first "site of loss", such as fat tissue, and be unable to move swiftly from this site to reach their target. As a result, the graph of BA against log P has a parabolic shape, with a maximum at which the best log P results in the highest level of activity.

The BA has usually two ways of behavior versus the lipophilicity of the compounds (Fig. 6.14), which can be assimilated to a straight line (6.9) in the most hydrophilic zone of the parabola or to a generic parabola that includes both the hydrophilic and hydrophobic zones (6.10).

$$\log(1/C) = k_1 \log P + k_2 \qquad \text{Linear equation} \qquad (6.9)$$

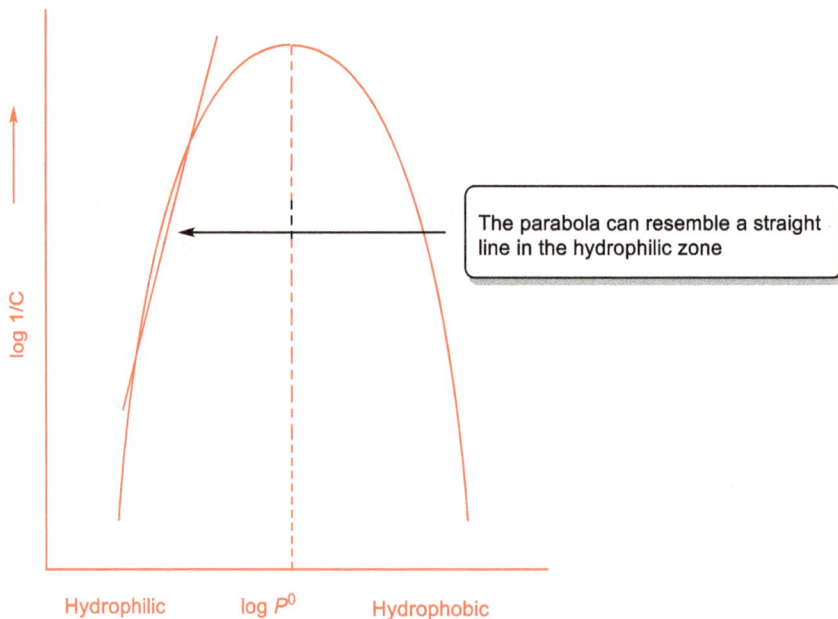

Fig. 6.14: Parabolic dependence of the biological response against the logarithm of the partition coefficient n-octanol/water.

$$\log(1/C) = k_1 (\log P)^2 + k_2 \log P + k_3 \quad \text{Parabolic equation} \quad (6.10)$$

Hansch and Fujita asserted that the contribution of a given substituent to the log P is a constant value. They defined the hydrophobic substituent constant by the following equation (6.11):

$$\pi_X = \log P_{RX}/P_{RH} \quad (6.11)$$

The more positive the π_X-value, the more lipophilic the substituent, and vice versa.

As an example, consider the log P values of benzene (log $P = 2.13$), chlorobenzene (log $P = 2.84$), and benzamide (log $P = 0.64$). Benzene is the reference compound, and the substituent constants of the Cl and the $CONH_2$ groups are 0.71 ($\pi_{Cl} = \log P_{Clbenzene} - \log P_{benzene}$) and -1.49 ($\pi_{CONH2} = \log P_{benzamide} - \log P_{benzene}$), respectively. Once these values are obtained, the theoretical log P value of m-chlorobenzamide can be calculated:

$$\log P_{m\text{-chlorobenzamide}} = \log P_{benzene} + \pi_{Cl} + \pi_{CONH_2} = 2.13 + 0.71 + (-1.49) = 1.35$$

The value observed of this compound is 1.35. It should be noted that the values of the aromatic substituents are different from those of the aliphatic substituents.

Let us calculate the log $P_{paracetamol}$, considering the three possibilities of calculation: from phenol, acetanilide, and benzene (Fig. 6.15).

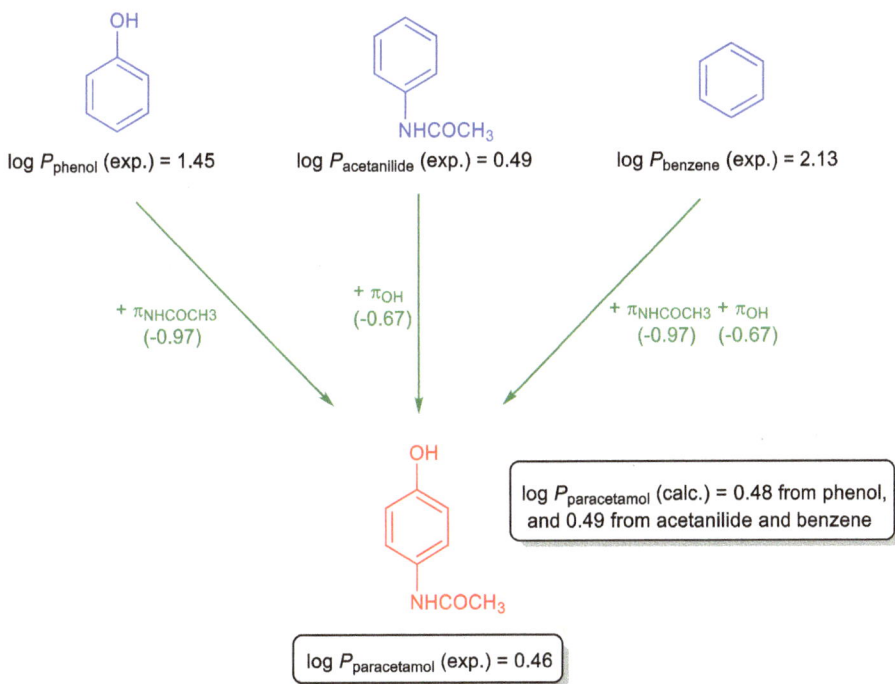

Fig. 6.15: Prediction of log P of paracetamol. See Tab. 6.1 for values of $\pi_{NHCOCH3}$ and π_{OH}.

Hansch equations can be written as follows:

$$\log\,(1/C) = -k_1(\log P)^2 + k_2\log P + k_3 \tag{6.12}$$

$$\log\,(1/C) = -k_4\pi^2 + k_5\pi + k_6 \tag{6.13}$$

where C is the concentration of the drug that produces a certain effect. The terms k_i are regression coefficients derived from the statistical treatment of the curves by the method of least squares (or least squares fitting).

6.7 Steric parameters

The steric characteristics of a molecule are intimately related to the ability of a molecule to bind to its receptor and, thus, to elicit its biological response (BR):

1. The steric parameter of Taft (E_s) has been derived by studying the acid-catalyzed hydrolysis of aliphatic esters:

$$E_s = \log k_x/k_0 \tag{6.14}$$

where k_x is the rate constant of the substituted compound and k_0 is the rate constant of the methyl ester.

2. Another measure of the steric factor is given by the molar refractivity (MR):

$$MR = (n^2 - 1)/(n^2 + 2) \times MW/d \tag{6.15}$$

where n is the refractive index, MW is the molecular weight, and d is the density. The term MW/d defines a volume, whereas $(n^2 - 1)/(n^2 + 2)$ is a corrective term.

Table 6.1 shows the electronic, hydrophobic, and steric descriptors of a number of substituents.

6.8 Craig plot

Although there are tables with the values of σ and π for a large set of substituents, it is often easier to visualize them using the Craig plot (Fig. 6.16).

It is possible to see at a glance which substituents have positive values of π- and σ-parameters, which have negative values and which have one positive and the other negative values. It is easy to see which substituents have similar values of π; for example, the ethyl, bromo, trifluoromethyl, and trifluoromethylsulfonyl groups are approximately in the same vertical of the representation. Therefore, these groups could be interchanged in drugs in which the main factor affecting BA is π. Similarly, groups in the same horizontal are isoelectronic or have similar values of σ (e.g. CO_2H, Cl, Br, and I).

Tab. 6.1: Electronic, hydrophobic, and steric descriptors.

Substituent	Aromatic π	Aliphatic π	σ_m	σ_p	MR^a	E_s
H	0.00	0.00	0.00	0.00	1.03	0.00
Br	0.86	0.60	0.39	0.23	8.88	−1.16
Cl	0.71	0.39	0.37	0.23	6.03	−0.97
F	0.14	−0.17	0.34	0.06	0.92	−0.46
I	1.12	1.00	0.35	0.18	13.94	−1.40
NO_2	−0.28	−0.85	0.71	0.78	7.36	−2.52
NMe_2	0.18	−0.30	−0.15	−0.83	15.55	
$^+NMe_3$	−5.96	−5.26	0.88	0.82		−2.84
NHMe	−0.47	−0.67	−0.30	−0.84	10.33	
NH_2	−1.23	−1.19	−0.16	−0.66	5.42	−0.61
$NHCOCH_3$	−0.97		0.21	0.00	14.93	
O^-	−3.87		−0.47	−0.81		
OH	−0.67	−1.12	0.12	−0.37	2.85	−1.12
OCH_3	−0.02		0.12	−0.27	7.87	−0.55
OEt	0.38	0.03	0.10	−0.24	12.47	
CN	−0.57	−0.84	0.56	0.66	6.33	−0.51
CHO	−0.65		0.35	0.42	6.88	
CO_2H	−0.32		0.37	0.45	6.93	
CF_3	0.88		0.43	0.54	5.02	−2.40
CH_3	0.56	0.50	−0.07	−0.17	5.65	−1.24
CH_2OH	−1.03		0.00	0.00	7.19	−1.21
$COCH_3$	−0.55		0.38	0.50	11.18	
C_6H_5	1.96	2.15	0.06	−0.01	25.36	−3.79
SO_2CH_3	−1.63		0.60	0.72	13,49	

^{a}The value of MR is scaled by a factor of 0.1.

6.9 Hansch equation

The Hansch analysis correlates BA with physicochemical properties by linear regression analysis, multiple linear regression, or nonlinear regression. Since practically all the parameters used in the Hansch analysis are linear values related to free energy (e.g. derived from velocity or equilibrium constants), the terms "linear approach related to free energy" or "extra thermodynamic approach" are used, as synonyms of Hansch's analysis. In its simplest version, the general form of the Hansch equation is given as follows:

$$\log(1/C) = k_1\sigma - k_2(\log P)^2 + k_3 \log P + k_4 E_s + k_5 \tag{6.16}$$

$$\log(1/C) = k_6\sigma - k_7\pi^2 + k_8\pi + k_9 E_s + k_{10} \tag{6.17}$$

The more active the drug, the smaller the concentration required, and the larger the value of $1/C$.

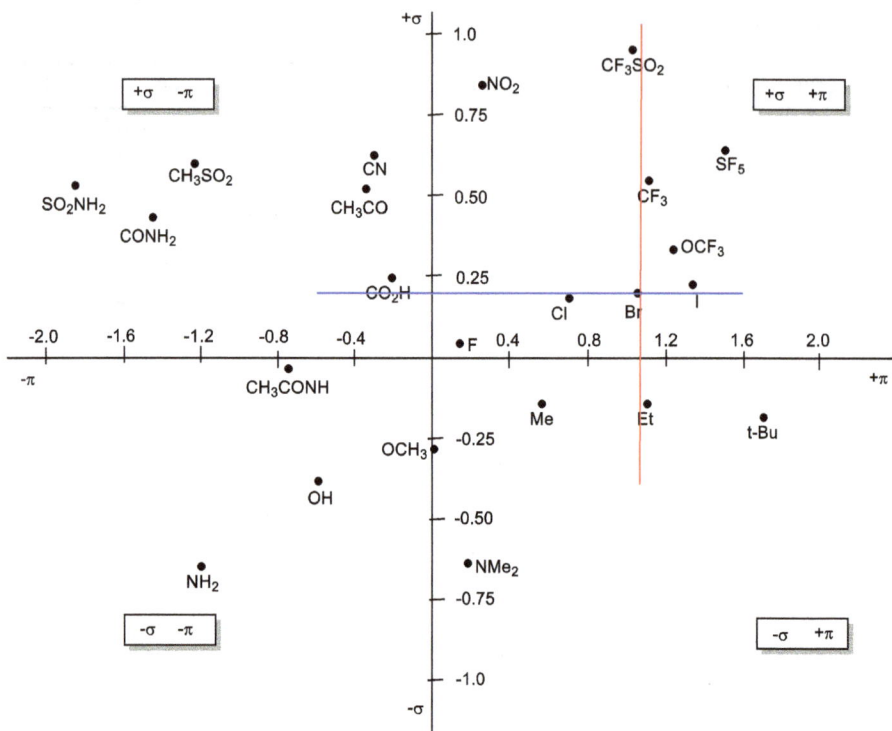

Fig. 6.16: Craig plot with the σ- and π-values of several substituents.

The accuracy of the adjustment of the equation to the experimental data can be estimated through the correlation coefficient (r^2). The fit is perfect when $r^2 = 1$. It is assumed that the equation is correct when $r = 0.9$ (or the same, $r^2 = 0.81$). The parameters selected for the "best equation" must be independent (i.e. the coefficient r should not be greater than 0.6–0.7; exceptions to this rule are combinations of linear and quadratic terms, such as log P and (log $P)^2$ that are normally highly correlated, with values of $r > 0.9$). Perhaps the most important consequence of Hansch's equations is the subsequent interpretation in physical–chemical terms of BA, which entails a better understanding of what happens at the molecular level.

With the equation obtained, in addition to a better understanding of the interactions, it will be possible *to predict* the structure of new compounds from the same family for which an optimum activity would be expected.

The number of compounds required to define a Hansch equation is a function of the number of variables that are to be introduced in the equation. As a rule of thumb, it is accepted that for each independent variable, at least five compounds are necessary.

6.10 Concepts and QSAR philosophy

– QSAR models are free-energy-related calculations:

$$\Delta G = -2.303 \, RT \log K$$

– Additivity of substituent group contributions to BA follows from many applications of Hansch analysis.
– Activity is usually expressed in $\log[C]$ or $\log 1/C$. Why?
  – Biological data is often found to be skewed, and log transformation makes the data to a normal distribution.

"The purpose of computing is insight, not numbers"
Richard Hamming (1915–1998)

6.11 3D QSAR model

In recent years, a method known as three-dimensional (3D) QSAR has been developed, according to which the 3D properties of the molecules as a whole are considered rather than considering individual substituents. The philosophy of the 3D QSAR model lies around the assumption that the most important characteristics of molecules are their overall size and shape, in addition to their electronic properties.

The best-known model is CoMFA (acronym of **co**mparative **m**olecular **f**ield **a**nalysis) which assumes that drug–receptor interactions are noncovalent, and that changes in BA correlate with changes in the steric and/or the electrostatic fields of drugs.

Networks or grids are increasingly used to measure the molecular properties. There are a number of molecular properties that can be measured as fields. A field can be defined as the influence that a given property exerts on the space surrounding the molecule. Consider a magnet as an example: it creates a magnetic field around it, which is stronger in the area of space closest to it. The most frequently measured molecular fields are steric and electrostatic. They can be measured by placing the molecule in a 3D network. Next, a probe atom such as a proton or an sp^3-hybridized carbocation is placed at each point in the network and a computer program calculates the steric and electrostatic interactions between the probe atom and the molecule. With respect to the steric field, it will increase as the probe approaches the molecule, while in relation to the electrostatic field, there will be an attraction between the positively charged probe and the electronically rich zones of the molecule, and a repulsion interaction between the probe and electronically deficient zones of the molecule. The points of the network having the same steric energy value are connected by contour surfaces, which define a steric field. A similar process will be carried out to measure electrostatic interactions. It is also possible to measure the hydrophobic field using water molecules as a probe.

A frequent problem in drug design is to decide on what conformation the drug molecule is when it links to the binding site of the biological target (active conformation). This is particularly true for flexible molecules that can adopt a large number of conformations. One might think that the most stable conformation is the active one. However, it may be that the less stable conformation is the active one: binding interactions with the target can result in an energy gain that compensates for the energy required to adopt that conformation. The simplest way to identify an active conformation is by studying the X-ray crystal structure of the complex between the protein and the ligand (drug). The structure of the ligand itself can be obtained from the Cambridge Structural Database, while that of the protein–ligand complexes can be obtained from the Protein Data Bank. The protein–ligand complex can be downloaded and studied through molecular modeling programs, whereby the ligand conformation can be identified. However, not all proteins can easily crystalize, and so other methods that are beyond the scope of this text must be used.

It is essential in the 3D QSAR study that the molecules are all in their active conformation and that they are placed in the network in exactly the same way. In other words, they have to be correctly aligned, thereby identifying the pharmacophore.

6.12 CoMFA advantages over traditional QSAR

Some of the problems associated with traditional QSAR are as follows:
(a) Only molecules of similar structure can be studied.
(b) The validity of numerical descriptors may be in question. These descriptors are obtained by measuring reaction rates and equilibrium constants of model reactions, and their values are tabulated in the specific scientific literature. However, the separation of one property from another is not always possible during experimental measurements. For example, the steric parameter of Taft cannot be considered purely to be a steric factor, because the reaction rates measured and used to define it are also affected by electronic factors. On the other hand, the n-octanol/water partition coefficients used for the calculation of log P are affected by the donor and acceptor nature of the hydrogen bonds of the molecules.
(c) Descriptors of unusual substituents may not be known.
(d) It is necessary to synthesize a set of molecules in which the substituents are varied to study a particular property (e.g. hydrophobicity), and these syntheses may be difficult.

These problems can be avoided with CoMFA, which has the following advantages:

(a) Favorable and unfavorable interactions are represented graphically by 3D contour surfaces around the representative molecule. A graphical representation like this is easier to visualize than a mathematical equation.
(b) The properties of molecules are calculated by means of a computer program. There is no dependence on experimental or tabulated values. There is no need to restrict the study to molecules of similar structure. Molecules can be studied by CoMFA as long as they share the same pharmacophore and interact in the same way with the biological target.
(c) Graphical representations of favorable and unfavorable interactions allow pharmaceutical chemists to design new structures. For example, if the boundary surface exhibits a favorable steric effect at a particular site, this may imply that the binding site of the biological target has an additional space that allows the drug to be extended to improve the receptor–drug interaction.
(d) The 3D QSAR approach can be used without knowledge of the biological target.

6.13 Potential problems of CoMFA

Some of the problems associated with CoMFA are as follows:

(a) It is very important to know the active conformation of each of the studied molecules. Identification of the active conformation is easy on rigid structures such as steroids, but is more difficult on flexible molecules in which various rotations of the single bonds are possible. Therefore, it would be advisable to have a conformational biologically active restricted analog that can act as a guide for the active conformation. The most flexible molecules could then be built on the computer, with the conformation resembling that of the rigid analog. This may be very useful in proposing the most likely active conformations of the ligands, if the structure of the binding site of the biological target is known.
(b) 3D QSAR provides a summary of how structural changes in drugs affect BA, but it is very dangerous to attempt to go further. For example, a 3D QSAR model may indicate that increasing the size of the molecule in a given location increases the activity. It might be suggested that an accessible hydrophobic pocket exists that allows extra bonding interactions. On the other hand, it is possible that the additional steric increase causes the molecule to bind with a different orientation with respect to the other molecules included in the study, and that this can be the cause of increased activity.

6.14 Key notes: quantitative structure–activity relationships

Introduction: QSAR involves the derivation of a mathematical formula, which relates the BAs of a group of compounds to their physicochemical properties. Traditional QSAR is carried out on a range of analogs: A series of compounds could then be constructed in the computer, with a common skeleton but having different substituents.

Procedure: A QSAR equation is derived relating BA to several physical features. Three physical features are of particular importance: hydrophobicity, electronic factors, and steric factors.

Hydrophobicity of the molecule: The hydrophobicity of a molecule is measured by its log P value, where P is the partition coefficient. The partition coefficient is the relative solubility of the compound in octanol and water.

Substituent hydrophobicity constant: The electronic properties of aromatic substituents are measured as Hammett's substituent constants (σ). Substituents with positive σ values are electron withdrawing, whereas substituents with negative values are electron releasing.

The steric factor: MR is a measure of size calculated from a substituent´s molecular weight, index of refraction, and density.

Hansch equation: The Hansch equation is the name given to the QSAR equation and usually contains physical factors such as log P, π, σ, and MR. The substituents used to derive a Hansch equation must represent a good spread of values for each physical parameter.

Craig plot: Craig plot compares two physical properties for different substituents. They are used to identify which substituents are valid for the derivation of a Hansch equation including these properties.

Three-dimensional QSAR: 3D QSAR involves the calculation of steric and electronic fields around molecules. It is not restricted to compounds having the same skeleton. The effect of steric and electronic fields can be shown visually by means of contour maps.

6.15 The significance of QSAR in drug discovery

Place 10 substituents of the four open positions of an asymmetrically disubstituted benzene ring: The number of compounds required for synthesis is 10,000. Nowadays, drug development is much too expensive to be guided by trial and error. QSAR, molecular modeling, and protein crystallography are important and valuable tools in computer-assisted drug design. The aim of this chapter is to give an introduction to the

QSAR methodology for beginners. The classical QSAR methods still play an important role in drug design. QSAR methods are cheap and efficient tools to derive and prove hypotheses on structure–activity relationships in a quantitative manner, especially in those cases where the 3D structure of the biological target is not known.

6.16 Exercises

1. Knowing that the log $P_{benzene}$ = 2.13, determine the log P values of the following compounds (the experimentally observed values are indicated in parentheses):

   (a) *m*-Xylene (3.20)

   (b) Mesitylene (1,3,5-trimethylbenzene) (3.43)

   (c) Hexamethylbenzene (2.33)

   (d) 1,3-Dinitrobenzene (1.49)

   (e) 2,4-Dihydroxybenzoic acid (1.44)

   Notice the increasing difference between the calculated and observed values in cases (a)–(c), and try to give an explanation. Do you find any justification for the deviation between the experimental and the calculated values?

2. A correlation between the bacteriostatic activity of methicillin-related penicillins and the π and σ values can be established for the X substituents on the aromatic nucleus:

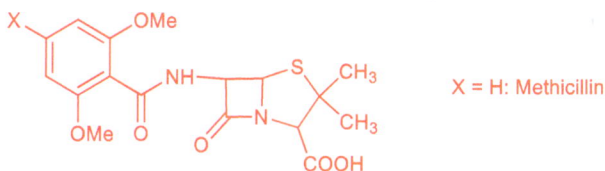

X = H: Methicillin

   The equation obtained for the bacteriostatic activity is

   $$\log (1/C) = -0.245\pi + 1.720\sigma + 1.776; \quad n = 10, \ r = 0.929$$

   (a) Comment on these two equations.

   (b) Propose different substituents to increase or decrease BA.

3. When studying the influence of substituent X on the activity of griseofulvin, the following QSAR relation was obtained:

   $$\log (1/\text{MIC}) = 0.56 \log P + 2.19\,\sigma X - 1.32; \quad n = 22, \ r = 0.93$$

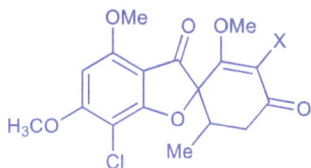

(a) What kind of substituents will give rise to maximum activity?
(b) Knowing that the antibiotic activity depends on an enone system that facilitates the nucleophilic attack to griseofulvin by the –SH groups, what relationship does it find between this mechanism and the previous equation?
Propose the attack reaction of the nucleophile mentioned above.

4. The adrenergic blocking activity of β-halo-β-arylethylamines can be expressed by the following equation:

$$\log\left(1/ED_{50}\right) = 1.22\,\pi - 1.59\,\sigma + 7.0; \quad n = 35, \quad r = 0.92$$

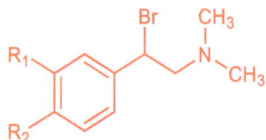

The compounds exhibit hepatic toxicity through a DNA alkylation mechanism represented by the equation

$$\log\left(1/ED_{50}\right) = 1.15\,\pi - 1.5\,\sigma^{+} + 7.89; \quad n = 22, \, r = 0.93$$

(a) Comment on these equations.
(b) Propose a DNA alkylation mechanism for this type of compound.
(c) Depending on the alkylation mechanism, explain why the parameter σ^{+} is used in the second equation.

5. The inhibitory activity of the monoamine oxidase produced by the benzylhydrazines $R\text{-}C_6H_4\text{-}CH_2\text{-}NH\text{-}NH_2$ in mice has been studied. For this, the necessary dose of drug before extracting their brain has been measured to produce a maximum response in the administration of L-DOPA to mice. The equation representing the inhibitory activity in vivo is

$$\log\left(1/C\right) = 0.304\,\pi - 0.183\,\pi^2 + 6.346; \quad n = 11, \, r = 0.856$$

(a) How and why does the in vivo inhibitory activity depend on the π-value of R?

(b) Of the following substituents, whose π-value is given in parentheses, what seems to be more interesting to increase the in vivo response?

H (0.0); Cl (0.71); Br (0.86); Me (0.56); *iso*-Pr (1.53); phenyl (1.96)

6. The in vitro antibacterial activity of sulfonamides **6.3** can be described by the following two equations:

$$\log (1/C) = -0.693 pK_a + 6.405; \quad n = 25, r = 0.962$$

$$\log (1/C) = 1.128\sigma + 0.398; \quad n = 25, r = 0.975$$

(a) What is the relationship between the two equations?

(b) It has been suggested that the active species is ionized sulfonamide, according to these equations. Why?

(c) What substituents are most suitable for the antibacterial activity?

(d) If antibacterial activity is studied in vivo, a parabolic dependence of the activity in relation to pK_a is observed. How can this observation be explained?

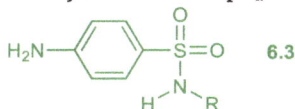

6.3

7. Pyrazoles **6.4** show inhibition of rat alcohol dehydrogenase hepatic, and may be utilized to prevent the conversion of methanol into formaldehyde and thus prevent methanol intoxication. The model relates the inhibition constant to the physico-chemical characteristics of substituents X. Analyze the characteristics of the model and as an example, calculate the log $1/K_i$ for a derivative **6.4** in which R = Cl:

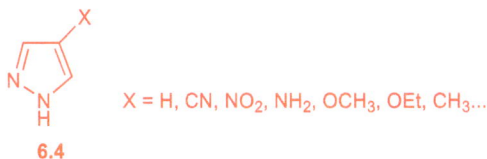

X = H, CN, NO$_2$, NH$_2$, OCH$_3$, OEt, CH$_3$...

6.4

$$\log (1/K_i) = 1.22 \, (\pm \, 0.16) \log P - 1.80 \, (\pm \, 0.78) \, \sigma_m + 4.87 \, (\pm \, 0.28);$$

$$n = 14, \, r^2 = 0.970, \, s = 0.316$$

$$c\log P_{\text{pyrazole}} = 0.28$$

s is the standard deviation from the regression equation. Its definition is explained in *Exploring QSAR. Fundamentals and Applications in Chemistry and Biology*, Hansch and Leo, found in the Fundamental bibliography of the Prologue of this volume.

8. The following QSAR equation was obtained for the antidepressant activity in humans, from a family of amino acid oxidase (MAO) inhibitors that respond to the chemical structure:

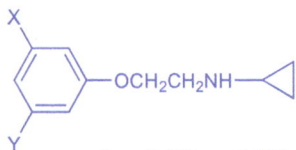

$$\log\ (1/C)\ =\ 0.398\,\pi + 1.089\,\sigma + 1.03\,E_s + 4.541;\ \ n = 9,\ r = 0.955$$

(a) Indicate on which parameters the inhibitory activity depends.
(b) Indicate the type of substituents that would be introduced in the aromatic ring to improve such an activity.
(c) Show two examples of substituents that would improve the activity.

9. The following QSAR was obtained for activity of the pesticide indicated in the structure. Explain the meanings of each term, and identify the type of substitute that could improve the activity:

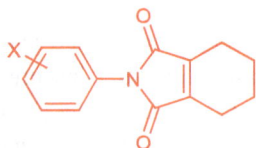

$$\log\ (1/C)\ =\ +\ 1.08\,\pi + 2.41\,\sigma + 5.25$$
$$n = 16,\ r = 0.84$$

10. The following reaction quantitatively relates the activity of an enzyme extracted from a lamb kidney with the MAO activity, caused by phenylglycines with different substituents in *meta* and *para*:

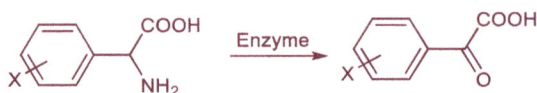

$$\log \ (1/K_m) = 0.3\pi + 0.6\sigma + 0.21E_s + 2.3$$
$$n = 5, \ r = 0.860$$

(a) Which parameters affect the enzyme's affinity of the substrate?
(b) How do these parameters have an influence, and what substituents should be introduced into the ring to increase the enzyme activity?

11. The following QSAR equation is related to the mutagenic activity of a series of nitrosamines. What kind of substituents will result in higher mutagenic activity?

$R_2N - NO$ $\log \ (1/C) = + 0.92 \, \pi + 2.08 \, \sigma - 3.26$

$N - $ Nitrosamines $n = 12, \ r = 0.794$

12. The muscarinic effects of a series of m-substituted benzyltrimethylammonium derivatives, expressed as BR, are indicated in the following equation:

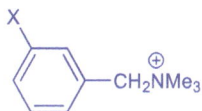

$$BR = 1.30 \, \pi - 0.41 \, E_s + 5.86$$
$$n = 10, \quad r = 0.90$$

(a) What parameters depend on this response?
(b) What substituents should be introduced to increase the activity?

13. Given the following QSAR equation, indicate the factors that would increase BA in the represented structure, and propose substituents that would improve such an activity:

$$\log\ (1/IC_{50}) = 0.481\,E_s + 0.606\,\pi\ + 4.81$$
$$n = 20,\ \ r = 0.93$$

14. The antifungal activity against *Cladosporium cucumerinum* from a series of aryl ethinyl sulfones is expressed by the following equation:

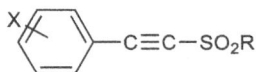

$$pIC_{50} = 1.10\,\sigma + 0.84\,\pi\ + 2.10\,E_s$$
$$n = 25,\ \ r = 0.89$$

(a) What parameters does it depend on?
(b) List some substituents to improve the activity.

15. The antimalarial activity of a series of aminoalcohols derived from phenanthrene is reflected in the following QSAR equation. With examples indicate the substituents that will favor BA.

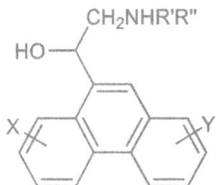

$$\log\ (1/C)\ =\ 0.27\,\pi_x\ +\ 0.40\,\pi_y\ +\ 0.65\,\sigma_x\ +\ 0.88\,\sigma_y\ +\ 2.34$$
$$n = 102,\ r = 0.913$$

7 Metabolic processes in drugs: other methodologies available for the discovery of new drugs

7.1 Goals

- To introduce the student into the concept of pharmacokinetics applied to the discovery of new drugs
- To introduce the student to the concept of prodrug as one of the solutions available to reduce the toxicity of the drug

7.2 Metabolism studies and their use in the discovery of new drugs

The action of drugs does not only depend on their capacity to develop a pharmacological response. It is also of great importance that they have pharmacokinetic properties that allow them to reach the site required for their action, and that their toxicity is minimal. Given the high degree of structural variability of drugs and the diversity of known metabolic reactions, it is necessary to establish relationships between their chemical structure and their pharmacokinetic properties: ADME (acronym for **a**bsorption, **d**istribution, **m**etabolization, and **e**xcretion).

The physicochemical properties of a drug will determine its ability to cross biological membranes, to deposit in fatty tissues, to bind to serum proteins, or to bind to its specific receptors to exert its action, and finally to undergo metabolic transformation and elimination.

I will detail the two most important steps in the development of new drugs that are absorption and metabolism.

7.3 Absorption

In any of the routes of administration, except intravenous, the drug has to cross biological membranes to reach its place of action. Since this is a critical step in the action of a drug, it must be taken into account. For example, in the gastrointestinal system, pH values vary from 1 to 3 in the stomach (due to the secretion of hydrochloric acid) to a value of 8 in the small intestine (ileum) and in the ascending colon. Therefore, the absorption of a drug is not equally effective in the different parts of the gastrointestinal system.

https://doi.org/10.1515/9783111316901-007

Neutral and lipid-soluble compounds are assured of their systemic action when administered orally, whereas the absorption of acids and bases depends on their dissociation constant (pK_a) and the pH of the medium, which are related according to the Henderson–Hasselbach equation:

$$pH = pK_a + \log [\text{basic form}]/[\text{acid form}]$$

Example: Aspirin® (Fig. 7.1).

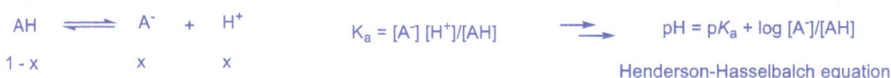

$$AH \rightleftharpoons A^- + H^+ \qquad K_a = [A^-][H^+]/[AH] \longrightarrow pH = pK_a + \log [A^-]/[AH]$$

$$1-x \qquad x \qquad x \qquad\qquad\qquad\qquad\qquad \text{Henderson-Hasselbalch equation}$$

where x is the degree of dissociation as decimal fraction

pK_a of Aspirine® is 3.6

Let us calculate how Aspirin® will be found at physiological pH (7.4):

$$7.4 = 3.6 + \log [A^-]/[AH]; \qquad [A^-]/[AH] = 10^{3.8} = 6310; \qquad x/1-x = 6310; \qquad x = 0.999\ 6\ 99.9\%$$

Therefore, at physiological pH Aspirin® will be fully ionized

However, in the stomach (pH = 1), Aspirin® is totally in the undissociated form:

$$[A^-]/[AH] = 10^{2.6} = 400; \qquad 1-x/x = 400; \qquad x = 2.10^{-3};\ 1-x \sim 1$$

Fig. 7.1: Calculation of the percentage of ionization of Aspirin® as a function of the pH of the medium.

A weak acid such as Aspirin® with a pK_a of 3.6 is practically in a nonionized form under the acidic conditions of the stomach, and for this reason it is rapidly absorbed. Once in the plasma (pH = 7.4), it ionizes almost completely and has no tendency to diffuse back into the stomach (Fig. 7.2).

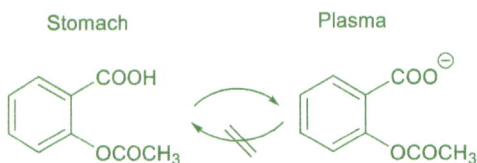

Stomach   Plasma

$pK_a = 3.6$

Once in plasma, it is ionized almost completely and has no tendency to diffuse to the stomach

Fig. 7.2: Forms in which Aspirin® is present in the stomach and plasma.

One of the objectives of pharmaceutical chemistry is to try to predict if a substance can be active, and if it can become a good drug. Although, of course, this is only known by synthesizing and testing it, Lipinski's rules can help us in predicting oral absorption.

7.4 Lipinski's rules (rule of 5)

Lipinski's rules emerged from a study of a wide-ranging oral drug group (WDI, Word Drug Index, more than 50,000 compounds), where its physicochemical properties were statistically correlated with its oral absorption profile, which in turn is related to its ability to cross membranes. It was observed that the drugs with the best oral absorption were those with a good balance between molecular weight (MW), lipid solubility and water solubility. These properties can be expressed quantitatively by means of four descriptors:
1. Solubility in lipids expressed by log P
   - Calculated log P (c log P) < 5 (Hansch)
   - Moriguchi log P (m log P) < 4.15
2. MW <500
3. H-bond donor groups <5
4. H-bond acceptor groups <10

Lipinski's rules are currently considered to be a good method to predict which structures with a pharmacological action will have good oral absorption. They are also known as the "rule of 5", because the parameter's cutoff values all contained 5s. In general, the more a compound is deviated from the Lipinski parameters, the lower the probability of its overcoming the more advanced stages of drug development (because it does not have a good oral absorption profile). Compounds that conform to these parameters are said to have drug properties. Despite the development of computational methods, none of these overcomes these simple rules in the ease of handling, nor are they used so widely.

Exercise: Apply Lipinski's rules to the drugs shown in Fig. 7.3.

7.5 Metabolism

We are now going to dedicate a broader section to the metabolism process, in which we will describe the metabolism of drugs from the chemical point of view, because we can design drugs whose metabolism is accelerated or reduced, but we have to know how they are metabolized. With the joint process of metabolism and elimination ends the action of drugs. Although metabolism is essentially a mechanism of de-

Aspirin ®

MW = 180 g/mol
mlog P = 1.70
H Acceptor groups = 4
H Donor groups = 1

Diazepam

MW = 284 g/mol
mlog P = 3.36
H Acceptor groups = 4
H Donor groups = 0

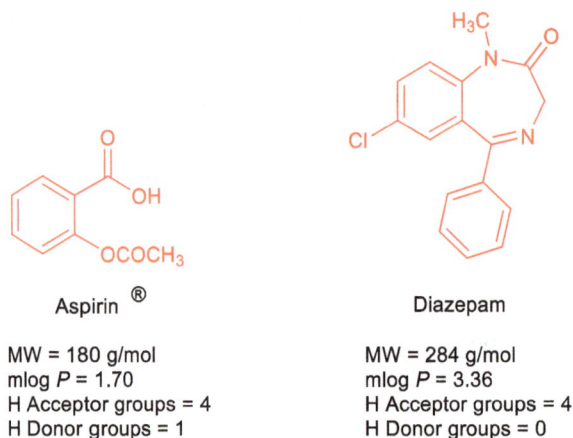

Fig. 7.3: Lipinski's rules applied to Aspirin® and diazepam.

toxification, in some cases, it produces the opposite, that is to say, toxicity of a drug, since toxic metabolites can be formed.

On the other hand, many compounds called prodrugs are inactive and are activated as drugs by metabolic processes, or it may happen that the metabolism causes changes in the pharmacological concentration of a compound. Therefore, the knowledge and prediction of metabolism is essential in the development of drugs.

Biotransformation of drugs can be classified into two broad groups:

(a) Phase I reactions directed to the formation of more water-soluble metabolites by unmasking or introduction of polar groups, such as COOH, OH, and NH_2.

Phase I reactions are catabolic (oxidation, reduction, and hydrolysis) and are usually products with a higher chemical reactivity and paradoxically may be more toxic or carcinogenic than before. Catabolic reactions are those where compounds are degraded to smaller molecules, e.g. glycolysis is the catabolic pathway of the carbohydrate degradation in the body. On the other hand, anabolic reactions are those where new molecules are synthesized from precursors. For example, protein synthesis is an anabolic process, because proteins are generated from amino acids (precursors).

(b) Phase II reactions complete the above with the formation of conjugates with sulfuric acid, glucuronic acid, or amino acids with existing polar groups or created in Phase I, thus forming more water-soluble species that are eliminated by the renal route. Phase II reactions are anabolic and often give inactive products (with some exceptions such as minoxidil sulfate). Reaction of Phases I and II occurs mostly in the liver.

7.5.1 Phase I reactions

7.5.1.1 Oxidation reactions

The main catalysts of the oxidation reactions are the cytochrome P450, which are hemoproteins. The oxidation process of an RH drug is schematized in Fig. 7.4.

$$RH + O_2 + \quad \underset{\text{(Reduced cofactor)}}{NADPH + H^+} \quad \xrightarrow{\text{P-450}} \quad ROH + H_2O + \quad \underset{\text{(Oxidated cofactor)}}{NADP^+}$$

Fig. 7.4: Oxidation of an RH drug by P450.

Nicotinamide adenine dinucleotide in its reduced form (NADPH) provides the system with one proton and two electrons, transforming it into the pyridinium derivative $NADP^+$ (Fig. 7.5).

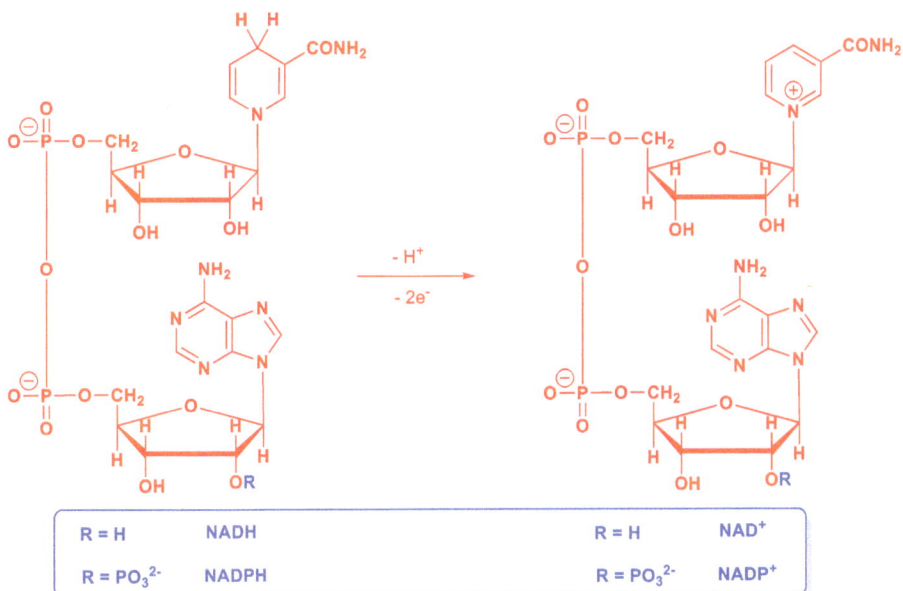

Fig. 7.5: Coupled transformation of NADH and NADPH cofactors in the oxidation reactions catalyzed by cytochrome P450.

These enzymes have the heme group as cofactor, a porphyrin containing Fe^{3+}, which in its reactive form of Fe^{5+} or Fe^{4+}, performs essentially hydroxylation and epoxidation reactions. The Fe^{3+} hydroperoxide is species **7.1**, where N represents the four pyrrole nitrogen atoms, and the fifth ligand is a cysteine residue of the enzyme. This species is converted by protonation and dehydration into Fe^{5+} (**7.2**) species, which is in resonance with Fe^{4+} (**7.3**). This active species breaks a C–H bond homolytically to give an alkyl radical and species **7.4** which, by radical coupling, originate alcohols and regenerate the heme group. On the other hand, the alkenes and arenes undergo the addition of radical 7 and, later on, they become epoxides. Scheme 7.1 shows the processes in a very simplified way; Silverman's text, within the fundamental bibliography in the Prologue of this volume, shows the complex real process.

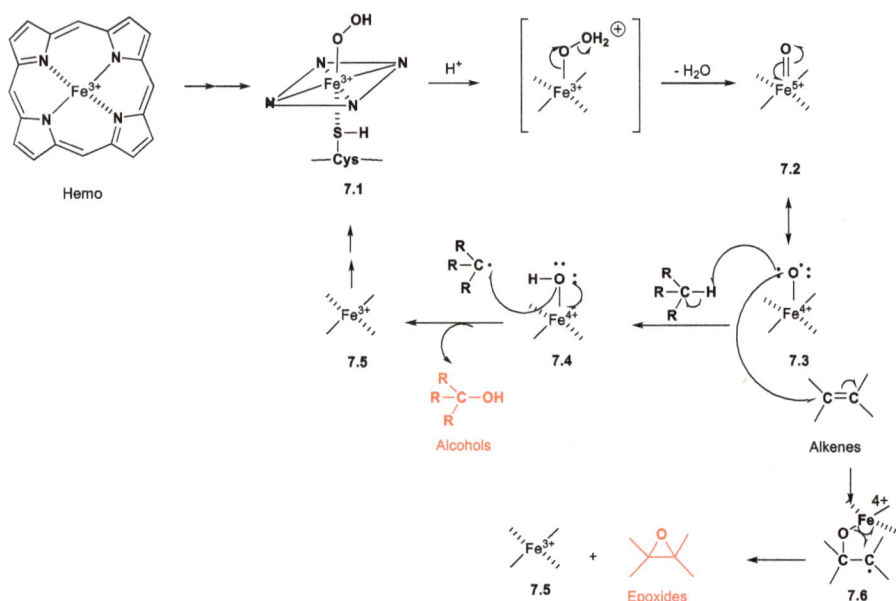

Scheme 7.1: Mechanisms of oxidation reactions of alkanes, alkenes, and arenes catalyzed by cytochrome P450.

The ease of oxidation is parallel to that which would be expected by the action of the chemical reagents. Thus the aryl, benzyl, aromatic, alkyl, oxidation of amines, and O-, S-, and N-dealkylation positions are oxidized:

(1) α-Positions of the carbonyl undergo oxidation.
(2) The aromatic oxidations take place in the activated position of the rings against the electrophilic attack. The presence of electron donor groups favors it, while a ring with electron-withdrawing groups is deactivated and its oxidation is difficult. When there is more than one ring, oxidation takes place preferably over the more activated one and in *para*-position (because it is the most activated locus and with less steric hindrance).

Example: Diazepam (Fig. 7.6).

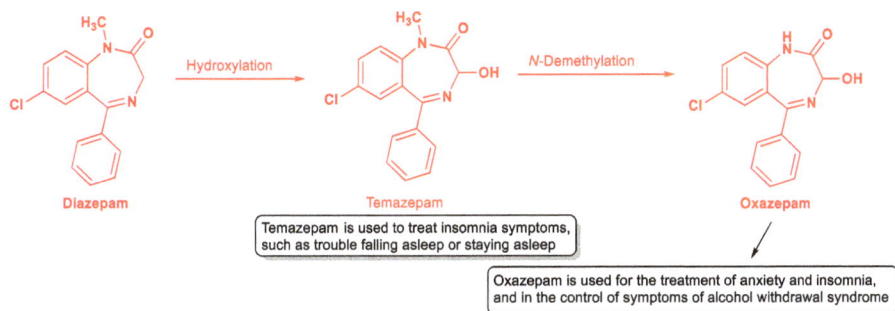

Fig. 7.6: Metabolism of diazepam.

The O-, S-, and N-dealkylations are hydroxylation reactions occurring on the α-carbon of the heteroatom and are followed by hydrolysis.

The proposed simplified mechanism is shown in Scheme 7.2.

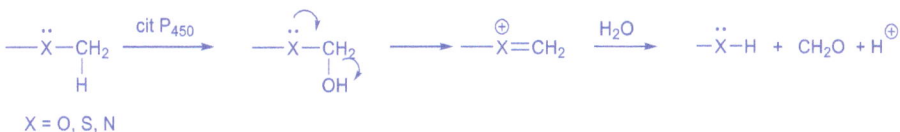

Scheme 7.2: Metabolism of O-, S-, and N-dealkylations.

Example: Phenacetin (Fig. 7.7).

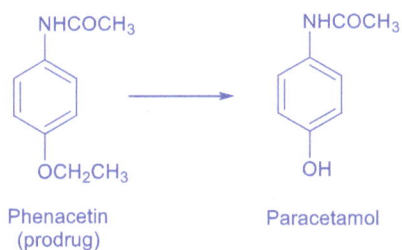

Fig. 7.7: Metabolism of phenacetin.

There are other metabolic oxidation processes, such as the oxidative deamination of dopamine, catalyzed by enzymatic systems other than cytochrome P450 (Scheme 7.3).

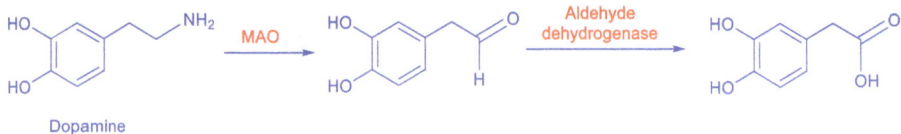

Scheme 7.3: Oxidative deamination of dopamine.

The most frequent reaction of the primary amines is the oxidative deamination by the action of monoaminoxidase in the unsubstituted amines in α.

7.5.1.2 Reduction reactions

Although the main metabolic pathway of drugs in mammals is oxidation, certain compounds whose functional groups are the azo, nitro, and carbonyl groups tend to be biotransformed by reduction to other functional groups such as amino and hydroxyl, directly susceptible to conjugation.

Example: Methadone (Fig. 7.8).

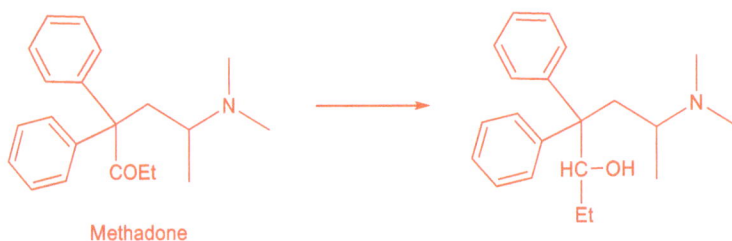

Fig. 7.8: Reductive metabolism of methadone.

7.5.1.3 Hydrolysis reactions

It is the immediate form of metabolism of esters and amides and takes place by the action of esterases and amidases that are very widespread in the organism.

Example: Procaine (Scheme 7.4).

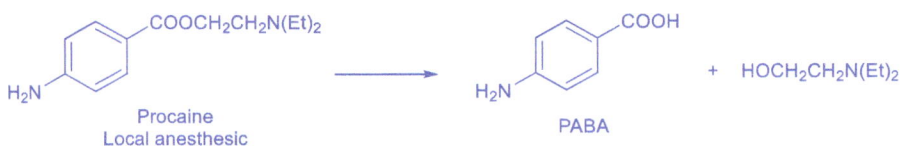

Scheme 7.4: Metabolic hydrolysis of procaine.

The amides hydrolyze more slowly than the esters; e.g. procainamide is more stable than procaine against hydrolysis (longer half-life) (Scheme 7.5).

Procainamide
Local anesthesic

PABA

Scheme 7.5: Hydrolysis of procainamide.

7.5.2 Phase II reactions

7.5.2.1 Conjugation reactions

They occur when the metabolites resulting from Phase I processes are not sufficiently water-soluble to be eliminated by urine. The purpose is to form more hydrophilic and rapid renal elimination metabolites. They take place with endogenous compounds (glucuronic acid, sulfate, glutathione, and certain amino acids).

7.5.2.1.1 The anomeric effect

The reason to explain the anomeric effect is outlined in Scheme 7.6.

7.5.2.1.1.1 Acyclic and cyclic forms of glucose

Alcohols are added to the carbonyl group of aldehydes in a fast and reversible way giving hemiacetals; this can be done intramolecularly when alcohols and carbonyl groups are at appropriate distances. Therefore, monosaccharides in aqueous solution are in equilibrium between the cyclic or hemiacetal form and the acyclic one, although the equilibrium is usually shifted toward the cyclic form since this is more stable.

Cyclic forms can be represented by the Haworth projection and the chair conformation (Scheme 7.7).

The OH at C5 reacts with the carbonyl group to give rise to a six-membered ring because it is more stable than the seven-membered one, and consequently the Haworth projection is depicted in Scheme 7.8.

7.5.2.1.1.2 Acyclic and cyclic forms of D-glucose

Glucose is usually present in solid form as a closed pyran ring. In aqueous solution, on the other hand, it is an open chain to a small extent and is present predominantly as an α- or β-pyranose, which interconvert (Scheme 7.9).

7.5.2.1.1.3 Molecular orbitals

To construct molecular orbitals (MOs) we need to combine the atomic orbitals of atoms that make up the molecule. This approach is known as the linear combination of atomic orbitals. Atomic orbitals are wave functions (Scheme 7.10), and the different

A) CONFORMATIONS OF MONO-SUBSTITUTED CYCLOHEXANES

Ring inversion (flipping) of cyclohexane

equatorial

axial

This conformation is lower in energy

The axial conformer is destabilized by the repulsion between the axial group X and the two axial hydrogen atoms on the same side of the ring

In these cases, only steric interactions must be considered

B) THE ANOMERIC EFFECT

In general, any tetrahydropyran bearing **an electronegative substituent in the 2-position** will prefer that substituent to be axial. This is known as the **anomeric effect**

for

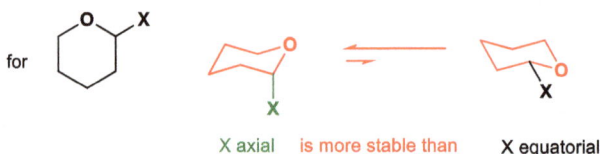

X axial   is more stable than   X equatorial

The whole explanation that I am going to give is to justify that in compound A, the phosphate group prefers to be axial: ELECTRONIC EFFECTS!!

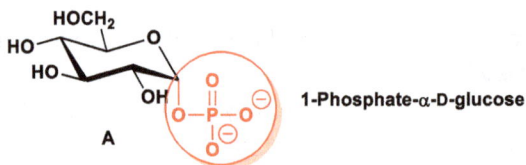

1-Phosphate-α-D-glucose

A

Scheme 7.6: The anomeric effect.

wave functions can be combined together in the way waves combine: they can add together constructively (in-phase) or destructively (out-of-phase).

Atomic orbitals can combine in the same way: in-phase or out-of-phase. Using two 1s orbitals drawn as circles with dots to mark the nuclei and *shading to represent phase*, we can combine them in-phase, that is, add them together, or out-of-phase when they cancel each other in a nodal plane in the center between the two nuclei. The resulting atoms are molecular rather than atomic orbitals (Scheme 7.11).

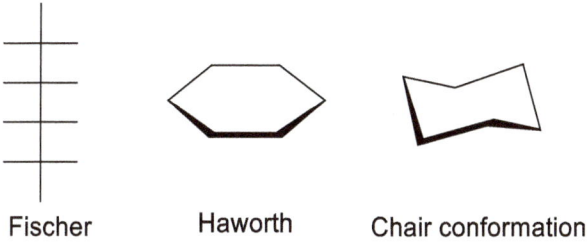

Fischer   Haworth   Chair conformation

Scheme 7.7: Representation of open and cyclic forms.

According to the general reactivity between aldehydes and alcohols, the formation of hemiacetals and acetals is possible:

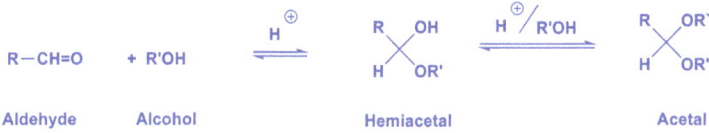

Aldehyde   Alcohol   Hemiacetal   Acetal

Nevertheless, the formation of a stable 6-membered cycle in sugars, stops at the formation of a hemiacetal:

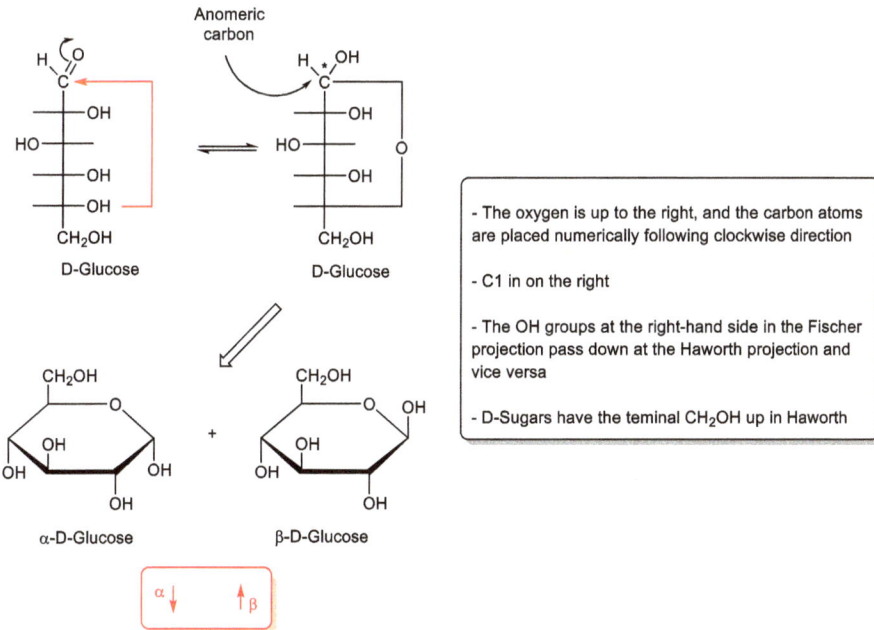

- The oxygen is up to the right, and the carbon atoms are placed numerically following clockwise direction

- C1 in on the right

- The OH groups at the right-hand side in the Fischer projection pass down at the Haworth projection and vice versa

- D-Sugars have the teminal CH_2OH up in Haworth

Scheme 7.8: Cyclic forms of D-glucose.

In the acyclic form of glucose C1 is achiral, but in the cyclic structure this carbon becomes chiral. The new center of chirality generated by the hemiacetal ring closure is called the anomeric center. The two stereoisomers are referred to as anomers

Although in principle two chair conformations are possible for the β-D-glucose, it is only important the one that possesses all the equatorial groups

The slow change of the optical rotation in solution is called mutarotation, and is interpreted by interconversion of the hemiacetals through the intermediate aldehyde. The equilibrium mixture contains 64% (β), 36% (α), and only 0.02% of free aldehyde

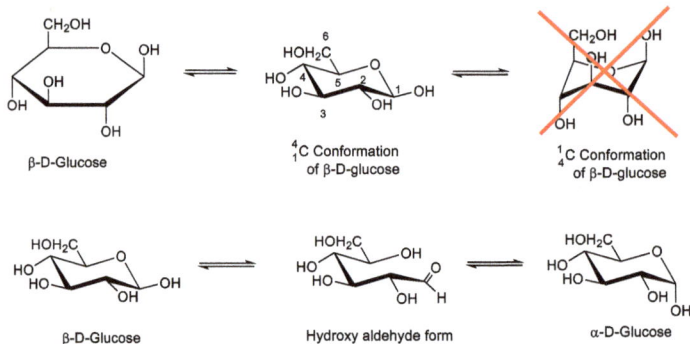

β-D-Glucose

$^4_1$C Conformation
of β-D-glucose

$^1_4$C Conformation
of β-D-glucose

β-D-Glucose

Hydroxy aldehyde form

α-D-Glucose

What is perhaps surprising is that the equatorial preference of the hydroxyl group located at the anomeric center is so small - only 2:1

Scheme 7.9: Cyclic and acyclic forms of D-glucose.

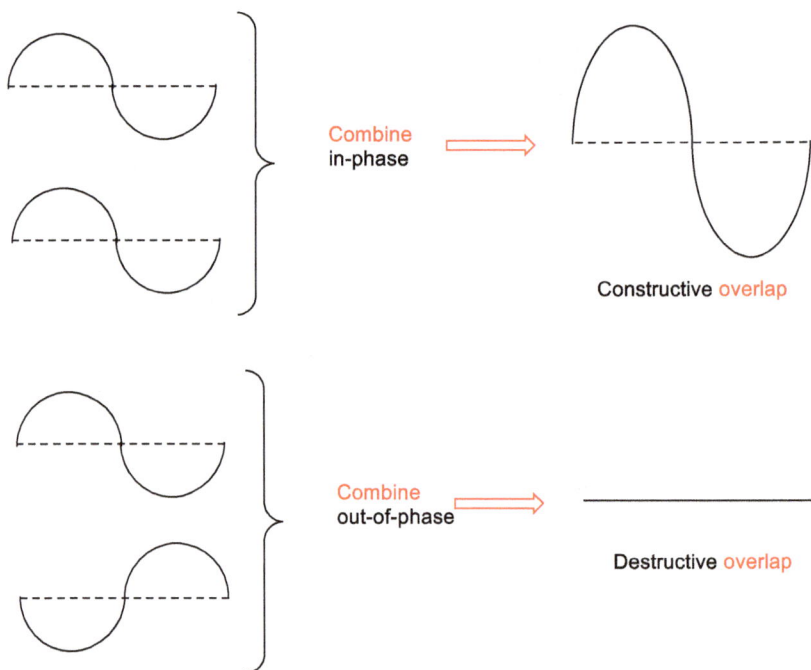

Combine in-phase

Constructive overlap

Combine out-of-phase

Destructive overlap

Scheme 7.10: Wave functions.

Nodal plane

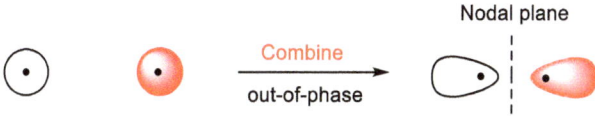

The two 1s orbitals combining out-of-phase to give an **antibonding orbital**

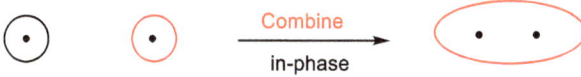

The two 1s orbitals combining in-phase to give a **bonding orbital**

Scheme 7.11: Molecular orbitals.

7.5.2.1.1.3.1 The hydrogen molecule

In the bonding MO, the electrons can be shared between the two nuclei, and this lowers their energy relative to the 1s atomic orbital. Electrons in the σ^* orbital do not help bond; in fact, they hinder the bonding (Scheme 7.12).

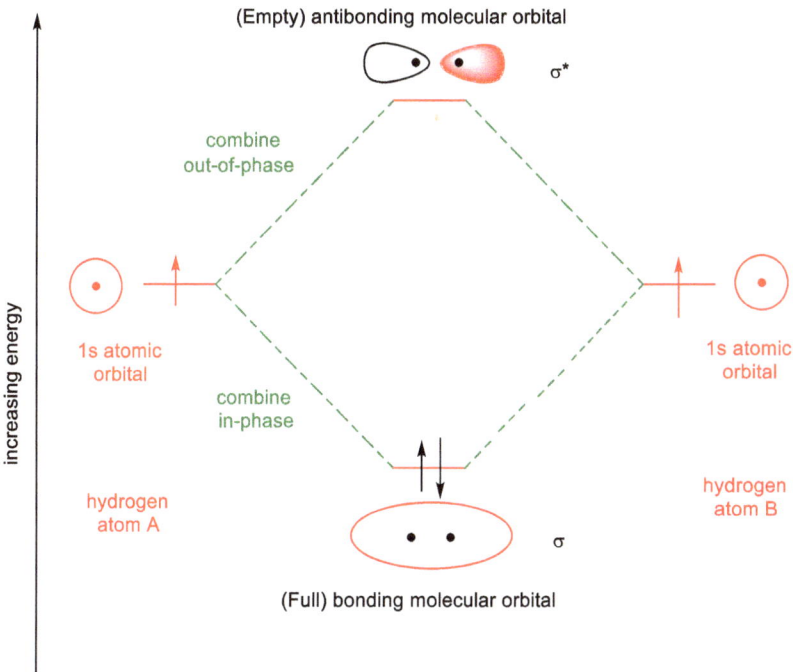

Scheme 7.12: The hydrogen molecule.

7.5.2.1.1.3.2 The anomeric effect

The **anomeric effect** is a stereoelectronic effect that describes the tendency of heteroatomic substituents adjacent to a heteroatom within a cyclohexane ring to prefer the *axial* orientation instead of the less hindered *equatorial* orientation that would be expected from steric considerations (Fig. 7.9).

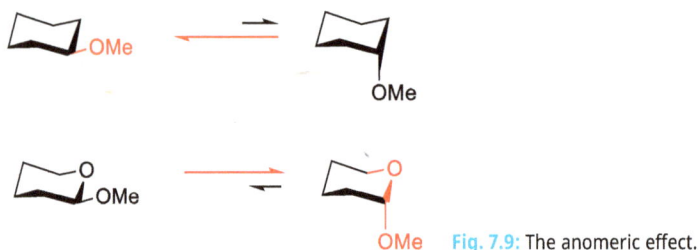

Fig. 7.9: The anomeric effect.

In the above case, the methoxy group on the cyclohexane ring (top) prefers the equatorial position. However, in the tetrahydropyran ring (bottom), the methoxy group prefers the axial position. This is because in the cyclohexane ring, the anomeric effect is not observed, and steric effects dominate the observed substituent position. In the tetrahydropyran ring, because of the endocyclic oxygen atom, the anomeric effect contributes and stabilizes the observed substituent position.

A widely accepted explanation is that there is a stabilizing interaction (hyperconjugation) between the unshared electron pair on the heteroatom (the endocyclic one in a sugar ring) and the σ^* orbital for the axial (exocyclic) C–X bond. This causes the molecule *to align the donating lone pair of electrons to the σ^* orbital* lowering the overall energy of the system and causing more stability. The fact that an antibonding orbital contributes to the destabilization of a molecule does not mean that an antibonding orbital is never occupied (Fig. 7.10).

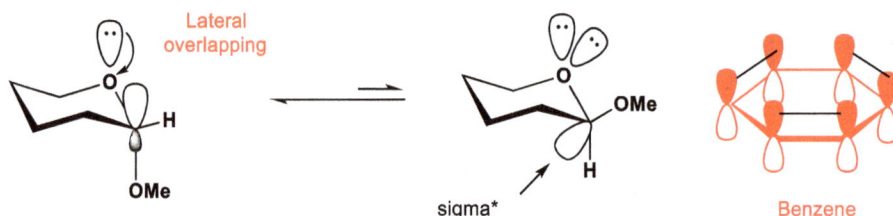

Fig. 7.10: The anomeric effect.

Compounds exhibiting an anomeric effect have a longer (and therefore weakened) bond outside the ring and a shorter, stronger C–O bond within the ring (Fig. 7.11).

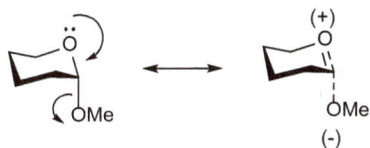

Fig. 7.11: One interpretation of the anomeric effect.

Another accepted explanation for the anomeric effect is the equatorial configuration that has the dipoles involving both heteroatoms partially aligned, and therefore repelling each other. By contrast, the axial configuration has these dipoles roughly opposing, thus representing a more stable and lower energy state (Fig. 7.12).

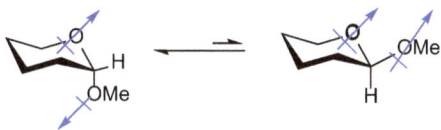

Fig. 7.12: A second interpretation of the anomeric effect.

7.5.2.2 Glucuronides

Conjugation reactions take place by the reaction of existing polar groups in a drug, and H_2SO_4, glucuronic acid, glutathione, sugars, or amino acids such as glycine, and also acetylation and methylation giving directly excretable compounds. Conjugation with glucuronic acid is probably the most important of all Phase II reactions. This is probably because there is a good supply of glucuronic acid in the body. Numerous alcohols, phenols, amines, thiols, and some carboxylic acids are metabolized by this pathway. The xenobiotic reacts with the activated form of glucuronic acid, glucuronic acid uridine diphosphate (UDPGA), to form a glucuronide conjugate very soluble in water (Scheme 7.13). The reaction is catalyzed by uridine diphosphate glucuronyl transferases (UDPG transferases).

The term "Walden inversion" is used to describe the stereochemical outcome of aliphatic bimolecular nucleophilic substitution reactions. A bimolecular nucleophilic substitution reaction *at a chiral carbon atom* produced a product that has the opposite stereochemistry from that of the reactant. This requires the nucleophile to approach the chiral atom from the side opposite to the leaving group (Scheme 7.13).

Example: Paracetamol conjugated with glucuronic acid (Scheme 7.14).

The reaction takes place on the anomeric carbon of the glucuronic acid giving rise to acetals.

7.5.2.3 Conjugation with glycine

Certain amino acids participate in the reaction of Phase II metabolites from aromatic carboxylic acids. Glycine is the amino acid that is commonly involved in these conjugations.

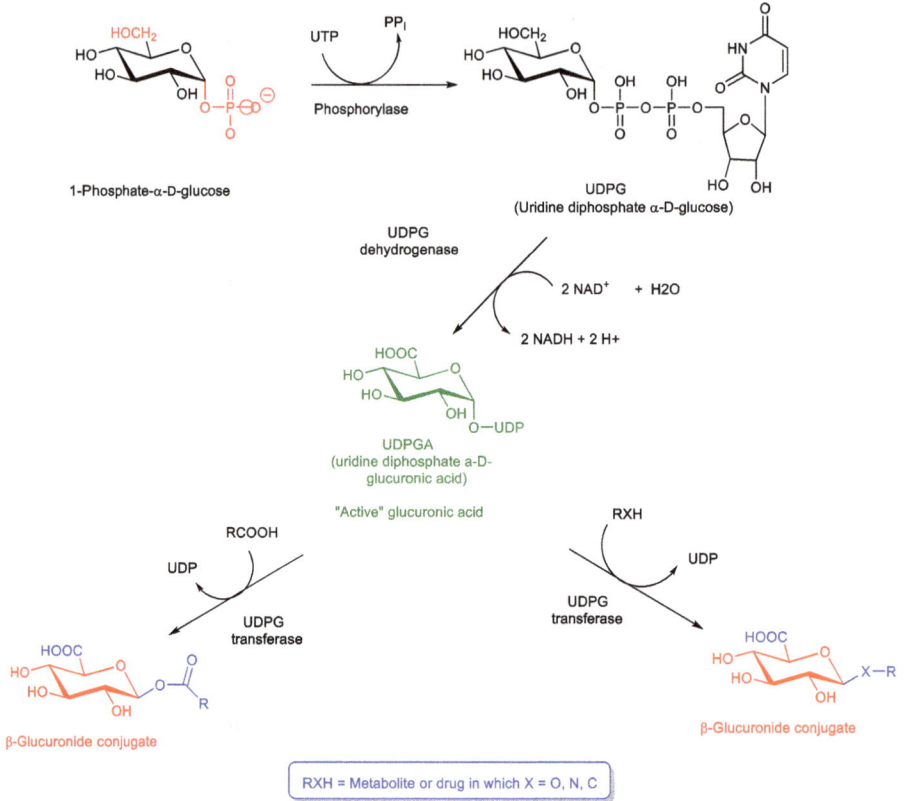

Scheme 7.13: Mechanism of formation of glucuronide conjugates. UTP is uridine triphosphate and X is O (alcohols and phenols), S (thiols), or NH (amines).

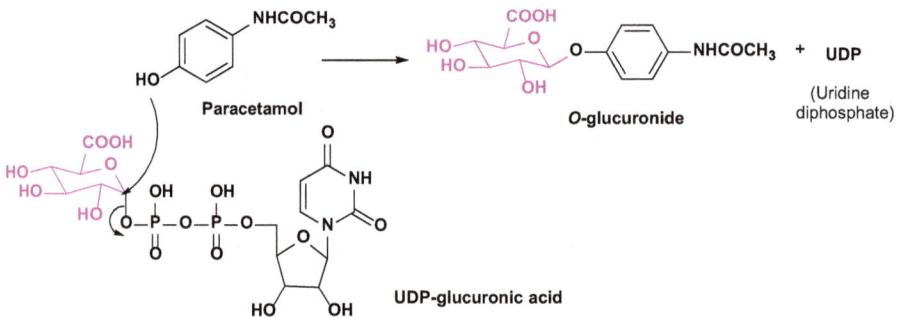

Scheme 7.14: Paracetamol conjugated with glucuronic acid.

Example: Scheme 7.15.

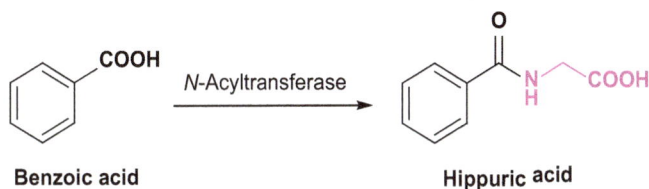

Benzoic acid   **Hippuric** acid

Scheme 7.15: Metabolization of benzoic acid.

Benzoic acid is introduced into the body through diet, as it is widely used as a preservative (both in its acid form and as a sodium, potassium, or calcium salt), although it is not a drug. Hippuric acid (from Greek: *Hyppos*, horse; *ouron*, urine) is an organic acid found in the urine of horses and other herbivores.

7.5.2.4 Acetylation

Acetylation is the main route of metabolization for amino groups and is mediated by acetyl-CoA.

Example: Acetylation of procainamide (Scheme 7.16).

Procainamide

Scheme 7.16: Acetylation of procainamide.

The half-life of *N*-acetylprocainamide is twice than that of procainamide and has no undesirable side effects.

Example: Metabolism of Aspirin® (Scheme 7.17).

7.5.2.5 Conjugation reactions with glutathione

Glutathione is a tripeptide containing a thiol group of great importance in the detoxification of drugs and xenobiotics. (A xenobiotic is a chemical substance found within an organism that is not naturally produced or expected to be present within the organism.) In the body, an equilibrium exists between the reduced form (GSH) and the oxidized form (GS–GS). Conjugation reactions of GSH are catalyzed by glutathione transferases. The conjugative reactivity of GSH is due to its thiol group (pK_a 9.0), which makes it a very effective nucleophile. The nucleophilic character is enhanced by deprotonation to a thiolate form. It reacts with halides, epoxides, or double activated bonds (Fig. 7.13).

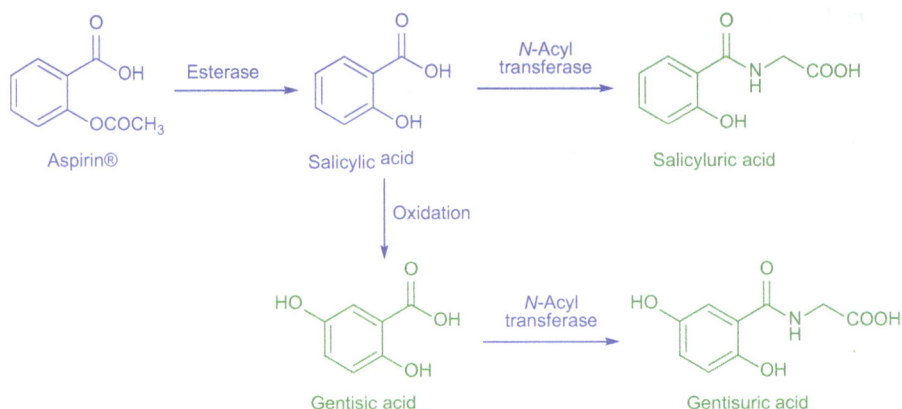

Scheme 7.17: Metabolism of Aspirin®.

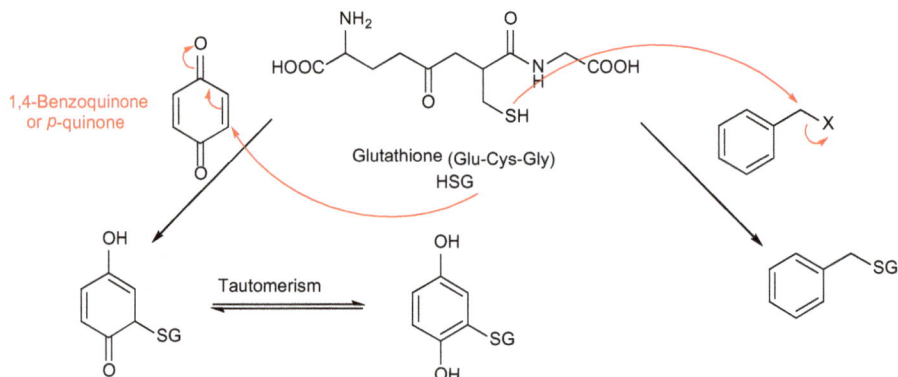

Fig. 7.13: Reaction of GSH with double activated bonds and benzyl halides.

GSH is added to α,β-unsaturated carbonyl compounds, a typical case in which the xenobiotic substrate is the toxic compound acrolein. The attack occurs on the activated CH_2 group. Quinones (*ortho* and *para*) and iminoquinones are strongly structurally related to α,β-unsaturated carbonyl compounds (Scheme 7.18).

Many *N*-hydroxylated products can be chemically unstable and dehydrated, producing electrophilic species of the imine or iminoquinone type, which may be toxic. The iminoquinone may also be produced through epoxide-type intermediates. Scheme 7.18 shows the transformation of paracetamol into an iminoquinone through a previous *N*-oxidation reaction. In Scheme 7.18, XH represents glutathione or a nucleophile present in endogenous macromolecules.

Large doses of acetaminophen cause liver and kidney damage in humans (the maximum recommended dose for an adult is 3–4 g/day at most). In therapeutic doses,

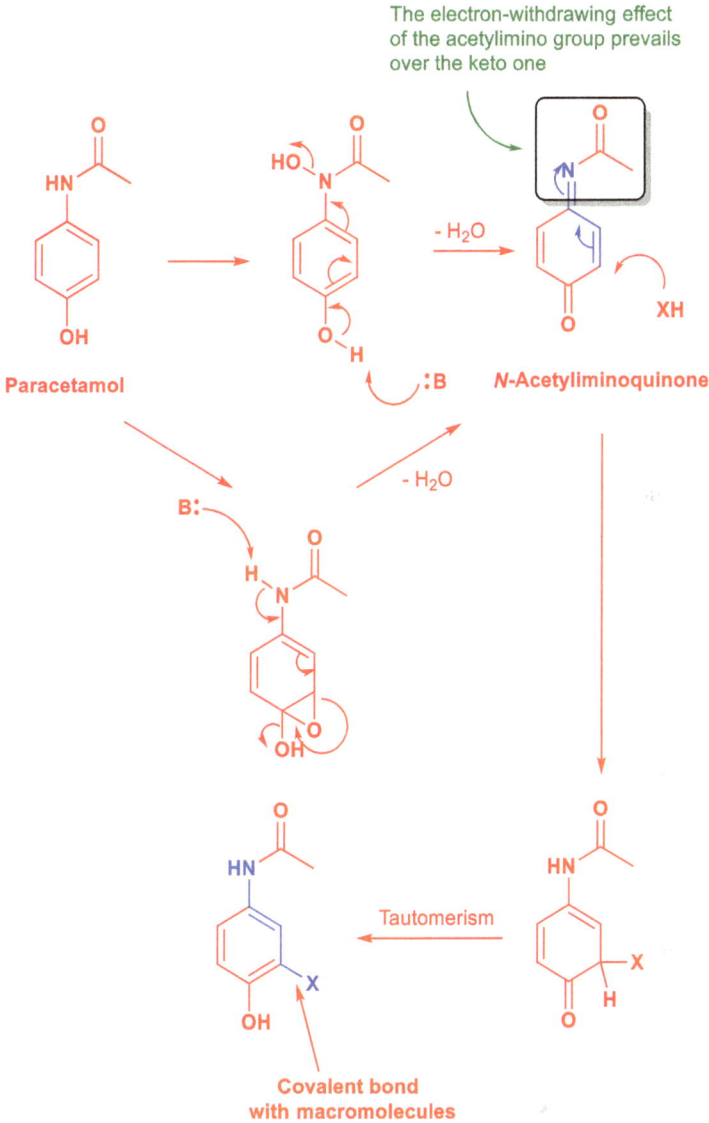

Scheme 7.18: Formation of iminoquinones (possible cytotoxic agents) in the oxidative metabolism of paracetamol.

acetaminophen is not very toxic, but in high doses it causes hepatic necrosis and renal lesions associated with a decrease in the glutathione reserves.

The treatment preferred for an overdose of paracetamol is the administration (usually in atomized form) of N-acetyl-L-cysteine, which is processed by the cells to L-cysteine, and used in *de novo* (*de novo* synthesis refers to the synthesis of complex molecules from simple molecules) synthesis of glutathione.

7.6 Prodrug concept

Prodrugs are inactive compounds, which give rise to a metabolite responsible for pharmacological activity. The design of prodrugs is usually carried out, with the purpose of modifying some pharmacokinetic or galenic characteristics of the drug in order to improve its therapeutic application. Here are some of the most frequent applications.

7.6.1 Improvements in the galenic formulations

Sometimes prodrugs are used due to issues of a galenic nature, such as increasing water solubility to prepare dosage forms with water. Thus, for example, prednisolone is a steroidal anti-inflammatory agent poorly soluble in water and therefore not administrable parenterally; its conversion into the corresponding hemisuccinate (hemi means "half") results in a water-soluble derivative, which can revert to the active drug by hydrolysis by the action of a plasma esterase (Fig. 7.14).

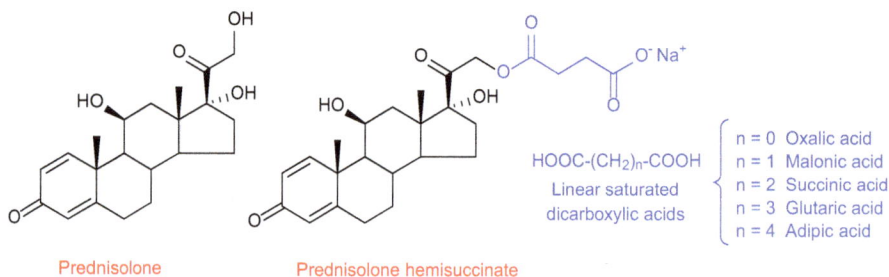

Prednisolone   Prednisolone hemisuccinate

$HOOC-(CH_2)_n-COOH$
Linear saturated
dicarboxylic acids

n = 0 Oxalic acid
n = 1 Malonic acid
n = 2 Succinic acid
n = 3 Glutaric acid
n = 4 Adipic acid

Fig. 7.14: Prednisolone hemisuccinate is a prodrug of prednisolone.

At other times, they are intended to improve the organoleptic characteristics in the pharmaceutical formulations. For example, chloramphenicol is an antibiotic of an intensely bitter taste that can be incorporated into syrups in the form of the corresponding tasteless palmitate (Fig. 7.15).

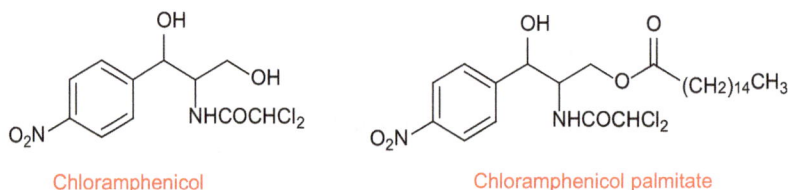

Chloramphenicol   Chloramphenicol palmitate

Fig. 7.15: Tasteless chloramphenicol palmitate.

7.6.2 Pharmacokinetic improvements

These can affect the release of the drug. For example, in hormones such as testosterone (an androgenic and anabolic hormone), a slow and constant release can be achieved through palmitate. In this way, the resulting prodrug is suitable for intramuscular administration at relatively high doses, accumulating in fatty tissues from which it will be released slowly by hydrolysis. This is the basis of the so-called depot action, which represents an important improvement both in the drug dosage regimens and in the resulting plasma levels (Fig. 7.16). Thus, very wide administrations that will allow the attainment of plasma levels of the hormone similar to their physiological levels will be possible.

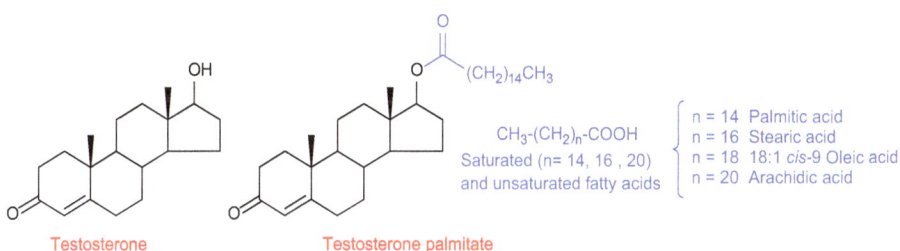

$$CH_3\text{-}(CH_2)_n\text{-}COOH$$
Saturated (n= 14, 16, 20)
and unsaturated fatty acids

n = 14 Palmitic acid
n = 16 Stearic acid
n = 18 18:1 *cis*-9 Oleic acid
n = 20 Arachidic acid

Testosterone   Testosterone palmitate

Fig. 7.16: Testosterone prodrug.

The molecules of all natural straight-chain fatty acids, saturated or unsaturated, contain an even number of carbon atoms.

Absorption may also be favorably modified by the use of prodrugs. This is the case of ampicillin, whose oral absorption is scarce due to its amphoteric character (it has $-NH_2$ and $-COOH$ groups, which ionize to $-NH_3^+$ and $-COO^-$ and is not well absorbed) (Fig. 7.17). Pivampicillin, a prodrug of ampicillin, is better absorbed than its parent drug.

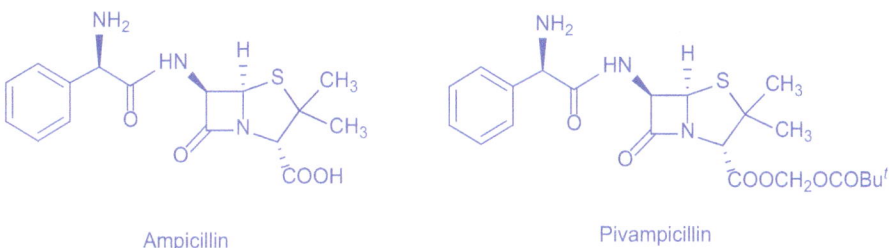

Ampicillin   Pivampicillin

Fig. 7.17: Pivampicillin, prodrug of ampicillin.

The use of prodrugs may also allow modification of the distribution. An example of this is found in antibacterial sulfonamides. These compounds have a polar group, which prevents their intestinal absorption. The metabolic elimination of this group by the intestinal bacterial flora leads to the active sulfonamide. As the release takes place in the final part of the gastrointestinal tract, its use is limited to the treatment of localized infections in this area. Phthalylsulfathiazole and sulfasalazine are representative examples of two of the most common strategies used in the design of these compounds. The former is a sulfathiazole pro-transporter in which the phthalyl group behaves as a labile modulating group, since it requires a hydrolytic process to provide the active species. On the other hand, sulfasalazine is an example of a bio-precursor prodrug, since it requires a non-hydrolytic activation process such as the reduction of the azo group (Scheme 7.19).

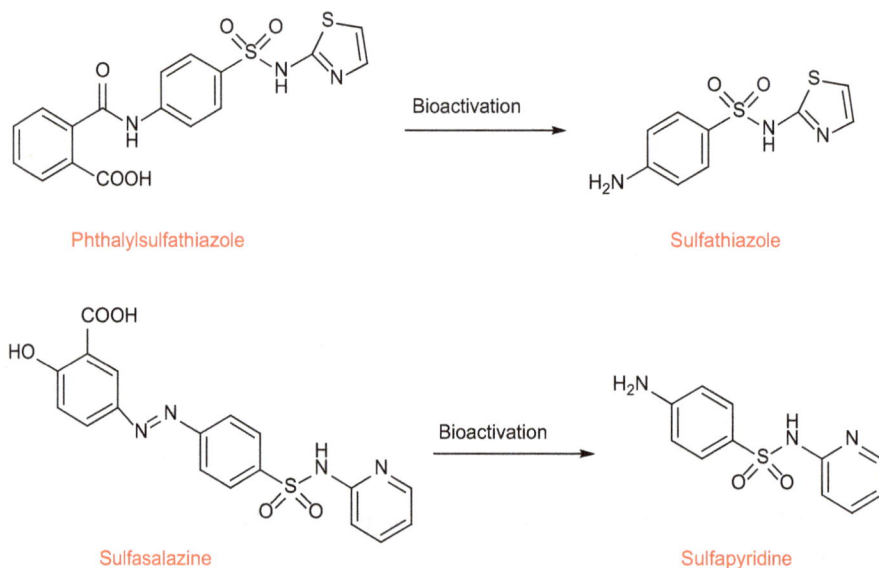

Scheme 7.19: Bioactivation of phthalylsulfathiazole and sulfasalazine.

7.7 Modulation of drug metabolism

The metabolic processes causing toxicity are often related to the oxidative reactions of Phase I, in which high reactivity intermediates such as epoxides or free radicals are generated, which interact easily with biomolecules inactivating them. To avoid this problem, different strategies can be adopted as follows:

(a) Completely suppress metabolic processes by administering drugs that are easily eliminated or that are resistant to these processes

(b) Focus metabolic reactions on parts of the structure that do not give rise to toxic compounds or facilitate other nonoxidative metabolic processes

7.8 Suppression of the metabolic processes

This strategy consists of the design of drugs that are stable against metabolic reactions. These types of substances are called hard drugs. They are very lipophilic compounds and tend to accumulate in the fatty tissues, producing long-term lesions. In addition, complete suppression of a drug metabolism is impossible in practice, because even if only a small percentage of metabolic reactions occur, these could produce toxic compounds. However, the principles on which hard drugs are based have served to prevent the first-pass effect of drug degradation in the liver before reaching the general circulation, as well as to prolong the action of drugs whose degradation is much faster than desirable.

One way to hinder the metabolism is to protect the group vulnerable to reaction through electronic or steric effects. For example, the hydroxylation of carbon directly attached to the nitrogen in the first *N*-dealkylation step of amines can be blocked by the presence of a bulky nitrogen-bonded group, which hinders the access of the enzyme. Hydroxylation of aromatic rings can also be blocked by the introduction of steric hindrance as well as by the presence of electron-withdrawing groups on the aromatic ring. A more robust group to this reaction may be substituted in order to avoid the hydrolysis of an ester group, such as an amide or by the introduction of a bulky group close to the position of the ester.

An example of such modifications is acetylcholine derivatives (Tab. 7.1) designed to extend the half-life of this neurotransmitter, the ester linkage, which is rapidly hydrolyzed by plasma esterases.

Tab. 7.1: Design of acetylcholine analogs.

R	R'	Hydrolysis velocity
H	H	500 (acetylcholine)
CH_3	H	15
CH_3	CH_3	0

7.9 Promotion of the nonoxidative metabolism

Drugs that are metabolized through a route that does not contain oxidative reactions are called soft drugs, designed to be inactivated in a controlled and predictable way. A soft drug can be defined as a biologically active compound, which is characterized by having a predictable and controllable in vivo destruction process (metabolism), preferably by means of a single hydrolytic process, to produce nontoxic metabolites, after having carried out its therapeutic role.

The easiest way to access this type of compounds is to incorporate groups vulnerable to hydrolysis or other nonoxidative reactions. One example is the muscle-relaxant decamethonium in which the substitution of two CH_2-CH_2 groups by two ester groups facilitates the hydrolysis reactions and avoids the toxicity accumulation problems of decamethonium (Fig. 7.18).

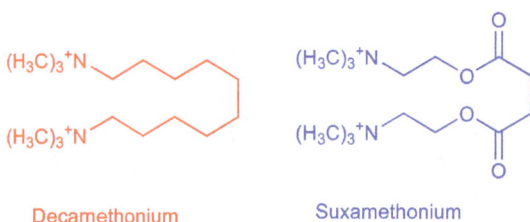

Decamethonium   Suxamethonium

Fig. 7.18: Hard (decamethonium) and soft (suxamethonium) drugs.

The concept of soft drug should not be confused with that of prodrug: while a prodrug is an inactive derivative, which is activated in vivo in a predictable active drug manner, a soft drug is an active species, but designed in such a way that it will undergo a predictable transformation or metabolism to an inactive metabolite. Accordingly, the common feature of prodrugs and soft drugs is that an in vivo transformation is involved in either activation (prodrug) or inactivation (soft drug). According to both definitions, the two concepts are opposed to each other.

The concept of soft drug can be very useful in the design of drugs with a better therapeutic index (TI). The fundamental goal of drug design is not only the search for molecules with a maximum biological activity. It is important to bear in mind that within a series of compounds, the optimum is not necessarily the one with the greatest activity, but the one with the highest TI. This parameter is a quantitative measure of the activity/toxicity ratio. For example, if the activity is quantified by means of the effective dose 50 (ED_{50}) and the toxicity as the lethal dose 50 (LD_{50}), since the values of ED_{50} and LD_{50} are in an inverse relationship with activity and toxicity, respectively, the TI is expressed in accordance with the following equation (7.1):

$$TI = activity/toxicity = LD_{50}/ED_{50} \qquad (7.1)$$

The optimization of a prototype or hit involves attempting to achieve the maximum TI, which can be achieved by lowering the denominator of the LD_{50}/ED_{50} fraction by increasing the activity, or by increasing the numerator by decreasing the toxicity.

The soft drug **7.1** is an isosteric analog of the antibacterial agent cetylpyridinium chloride **7.2**. Both compounds have the same length of the hydrophobic chain and have remarkable antimicrobial properties. This is because **7.1** can undergo an easy hydrolytic deactivation, thereby destroying the positive charge of the pyridinium ring (Scheme 7.20).

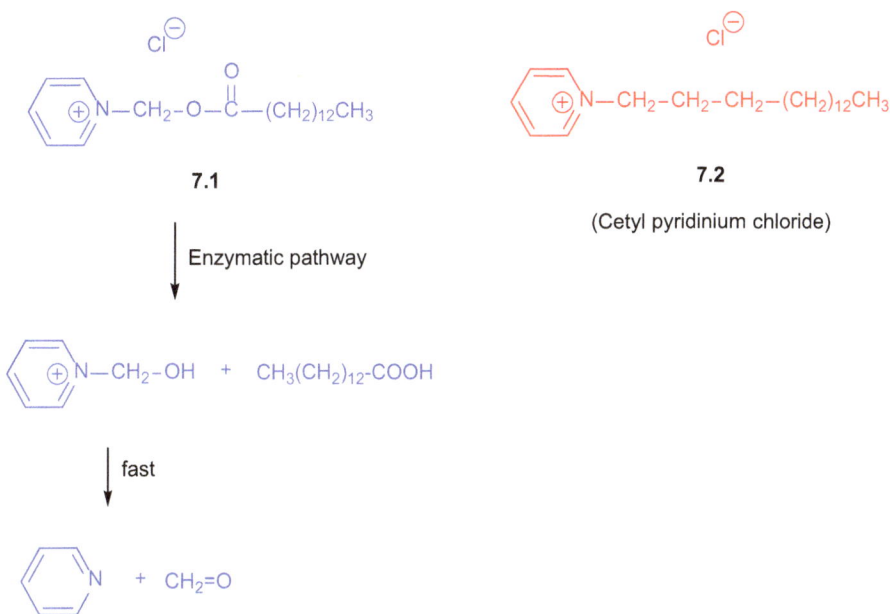

7.1

7.2

(Cetyl pyridinium chloride)

Enzymatic pathway

fast

Scheme 7.20: Soft analog of cetylpyridinium chloride and its metabolic transformation.

7.10 Summary

Drug interactions with various receptor systems and their biochemical and pharmacological effects were the focus of Chapter 5. The pharmacokinetic phase, which consists of the physicochemical processes that enable the drug to reach its site of action, comes before this pharmacodynamic phase in the timeline of pharmacological action. Any medicine to be dissolved must first be absorbed at the administration site before it can be distributed throughout the organism. However, because it depends on so many variables, this technique is very inaccurate in terms of the drug's ability to reach its many targets. A drug will bind nonspecifically, get stuck in depot sites, and travel to other systems in addition to the specific receptor binding that causes the desired action. These nonspecific bindings result in toxicity and unwanted side effects. The metabolism of the drug molecule must be taken into account along with the pharmacokinetic and pharmacodynamic stages of pharmacological activity. Numerous enzyme systems involved in the regular upkeep of cells are vulnerable to drugs. These systems are able to identify a "xenobiotic" molecule and expose it to biotransformation, frequently in unintended ways that aim to get rid of xenobiotic agents. Furthermore, the metabolites produced by such reactions might also have pharmacological effect. An inactive prodrug can also be activated by metabolism to become the active substance.

Drug's activity is ended by its excretion, either prior to or following biotransformation, just as crucial to comprehending the whole pharmacological activity of a medicine as molecular and biochemical pharmacodynamics are pharmacokinetics and drug metabolism. Activity in these sectors has significantly expanded as a result of methodology's generally quick advancement and the expectations of drug regulatory bodies. The proper consideration of the drug distribution and metabolism, as well as the manner in which their alteration might enhance the overall efficacy of pharmacological action, has also been acknowledged as being necessary for rational drug design.

7.11 Key notes: drug metabolism

Definition: Drug metabolism refers to the reactions undergone by a drug in the body. Metabolic enzymes exist mainly in the liver and catalyze reactions that increase the polarity of the drug.

Phase I metabolism: Phase I reactions usually involve the introduction of polar groups to a drug so that the drug can be excreted more easily. Particular groups in a molecule are more prone to metabolism than others. The cytochrome P450 enzymes are important metabolic enzymes involved in oxidative reactions. Drugs that enhance or inhibit the activity of these enzymes will affect the levels of drugs that are normally metabolized by them.

Phase II metabolism: Phase II reactions involve the formation of polar conjugates. Highly polar molecules such as glucose are linked to polar functional groups that may have been placed there by Phase I metabolism.

Anomeric effect: The **anomeric effect** refers to the *unexpected* preference for an electronegative substituent in the 2-position of a tetrahydropyran to be **axial**.

7.12 Exercises

1. Propose reasonable Phase I metabolic reactions that may occur in the following drugs:

(a) Apoatropine

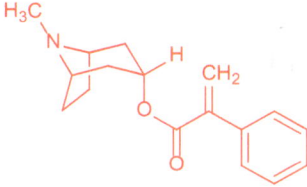

(b) Ampicillin

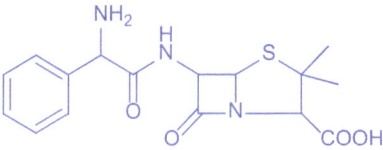

(c) Eseridine

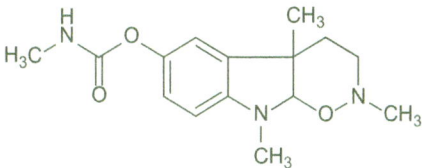

(d) Dibenzepine

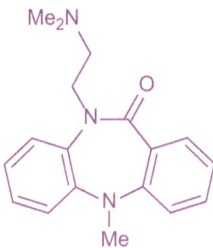

(e) Propoxyphene

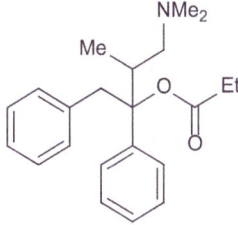

(f) Reboxetine

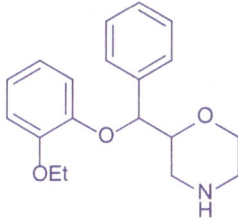

(g) Amitriptyline

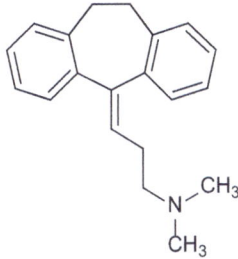

(h) Methylphenidate

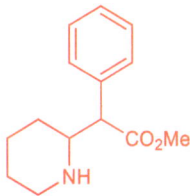

2. Explain how the following metabolic transformations occur, detailing the steps:
   (a) Methylphenobarbital [1-methyl-5-ethyl-5-phenylbarbituric acid] into 5-ethyl-5-(4-hydroxyphenyl)barbituric acid
   (b) 1-(p-Tolyl)-N-methyl-2-propanamine into [4-(2-aminopropyl)phenyl]methanol
   (c) Adrenaline: [1-(3,4-dihydroxyphenyl)-2-methylaminoethanol] into 1-(3,4-di-hydroxyphenyl)-2-aminoethanol, and this into 2-(3-methoxy-4-hydroxyphenyl)-2-hydroxyacetic acid
   (d) Indomethacin: [2-(2-methyl-5-methoxy-1-(p-chlorobenzoyl)-3-indolyl)acetic acid] into 2-(2-methyl-5-hydroxy-3-indolyl)acetic acid 5-glucuronide

3. Tripeptide glutathione (γ-glutamyl-cysteinyl-glycine) is widely distributed in mammalian tissues. One of its physiological functions is to react with all kinds of electrophilic species, thus avoiding the toxicity that would derive from the reaction of these with biomolecules.
   (a) Formulate the structure of glutathione.
   (b) Propose the mechanism by which the following compounds are reacted with glutathione:
   (a') N-(4-Chloromethyl thiazol-2-yl)acetamide
   (b') Busulfan

H$_3$CO$_2$SO⌒⌒⌒OSO$_2$CH$_3$

   (c') Arecoline

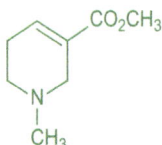

   (d') Crotonaldehyde

4. Classify the compounds on the right as "soft" or "hard" analogs of those on the left, according to the structural modification performed (D = drug):
   (a) D-N(CH$_3$)$_2$ D-N(CH$_3$)C(CH$_3$)$_3$
   (b) D-CH$_3$ D-COOCH$_3$
   (c) D-COOC$_2$H$_5$ D-COOC(CH$_3$)$_3$
   (d) D-C$_6$H$_5$ D-C$_6$H$_4$-pCF$_3$

(e) D-CH$_2$COOR D-CH$_2$CONHR

(f) D-COO(CH$_2$)$_2$N$^+$R$_3$ D-COOCH$_2$N$^+$R$_3$

5. Explain the modification made in the following drugs when preparing them as modified drugs:

	Drug	Modified drug
a)		
b)		
c)		
d)		
e)		

6. Atracurium was designed as a soft muscle relaxant in which two ester groups are introduced in the two β-positions with respect to the quaternary nitrogen atoms in the linker chain. Their presence makes it possible for a spontaneous elimination reaction to take place at physiological temperature and pH, aided by the acidity of neighboring carbonyl protons. Specify the removal mechanism that takes place:

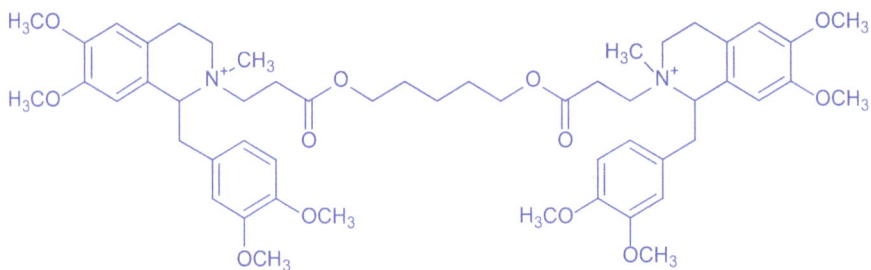

Atracurium

8 Synthetic drug strategies

8.1 Goals

- Knowledge of retrosynthetic analysis as a way to simplify the structure of the target molecule until reaching simple starting materials
- Ability to propose the synthesis phase, based on the information obtained in the retrosynthetic analysis, based on mechanisms of reaction and reactivity of organic compounds

8.2 Introduction

Chemistry is a difficult discipline and presents a difficulty similar to learning a second language. As with any foreign language, we must learn grammar rather than memorize phrases. We are concerned with the processes by which reactions occur, the "how and why" of the field. We get a command of organic chemistry based much more upon understanding the basics than upon memorization. We are going to use the concept of electron flow together with the rigorous use of curved arrow as an electron bookkeeping unit. The concept of flow is very important. Just as water flows under the influence of gravity, electrons flow under the influence of charge: from electron-rich atoms to electron-deficient atoms. Lewis dot structures are used to keep track of all electrons, and curved arrows are used to symbolize electron movement. You must be able to draw a proper Lewis structure complete with formal charges accurately and quickly. Your command of curved arrows also must be automatic.

To carry out an analytical approach to the design of organic synthesis, a basic knowledge of organic chemistry is required. The strategy followed is one of backward analyses, i.e. from products to reagents, which is called retrosynthetic analysis. The methodology consists of gradually "disarming" the product by breaking strategic bonds to give simpler fragments that we call building blocks or synthons. Bond-breaking is called a disconnection.

Suppose that we want to prepare barbital, the first barbiturate synthesized with a hypnotic-sedative activity. Its synthesis is carried out starting from diethyl malonate and ethyl bromide, in the presence of sodium ethoxide, by an S_N2 process, due to the acidity of the hydrogen atoms of the malonate C-2. It is then condensed with urea in a basic medium to give barbital (Scheme 8.1).

Suppose, however, that this barbiturate has not been synthesized before and that we must design its synthesis. In this case, the starting products are unknown, and all that is known is the structure of the target molecule (TM). It is obviously necessary to start with this structure and work backward. The key to the problem is the TM functional groups, which in our case will be nitrogen atoms, carbonyl groups at positions 4 and 6, and C-5.

https://doi.org/10.1515/9783111316901-008

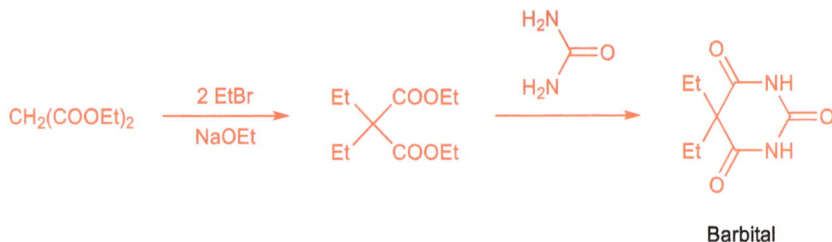

Scheme 8.1: Synthesis of barbital.

There are one or several valid disconnections for most functional groups, i.e. imaginative processes that reverse the actual chemical reactions by means of the rupture of a bond in the TM, to lead to a new compound, from which the TM can be synthesized. In this case, the first disconnection (a) will be that of a C–N bond, which leads to the precursors diethyl 2,2-diethylmalonate and urea (Scheme 8.2).

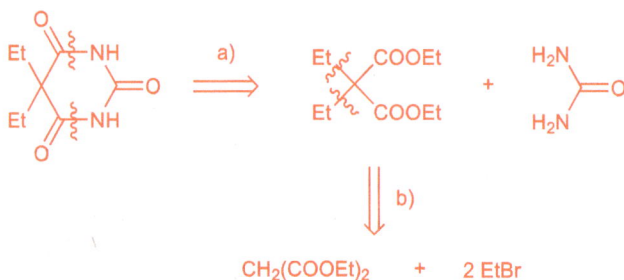

Scheme 8.2: Retrosynthesis of barbital.

The second (b), that of a C–C bond of a 1,3-dicarbonyl compound, leads to diethyl malonate and ethyl bromide. Alkylation of diethyl malonate proceeds well when primary halides are used, since it is an S_N2. When it is necessary to use secondary halides, it will be necessary to look for another alternative, as we will see later.

8.3 Definitions

Disconnection: The analytical operation that consists of the rupture of a bond, which converts the molecule into a starting product. It is the reverse operation of a chemical reaction. It is represented by a double arrow and a crossed wavy line perpendicular to the fragmenting bond.

FGI (functional group interconversion): It is the substitution of a functional group for another, which allows the realization of the disconnection. This is again the in-

verse of a chemical reaction and is represented with the same symbol of the disconnection and the FGI acronym on the double arrow.

Reagent: The compound whose reaction leads to an intermediate in the proposed synthesis, or leads to the TM. It is the synthetic equivalent of a synthon.

Synthetic equivalent: A reagent that plays the role of a synthon when it cannot be
used, usually due to its instability.

Synthon: A fragment (usually an ion), produced by a disconnection.

Target molecule: The molecule whose synthesis is intended. It is represented as TM,
followed by its corresponding number.

8.4 Rules to make a good disconnection

- A logical mechanism of reaction
- The greatest simplicity possible
- Leading to readily available starting materials
- Use FGI to make disconnections easier

As is well known, *tert*-butyl alcohol can be obtained by hydrolysis of *tert*-butyl chloride (Scheme 8.3).

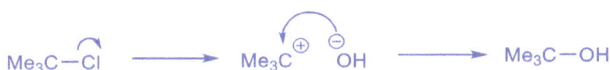

$$Me_3C-Cl \longrightarrow Me_3C^{\oplus} \quad ^{\ominus}OH \longrightarrow Me_3C-OH$$

Scheme 8.3: Synthesis of *tert*-butyl alcohol.

The mechanism for the inverse imaginary reaction is shown in Scheme 8.4.

$$Me_3C-OH \Longrightarrow Me_3C^{\oplus} \quad ^{\ominus}Cl \Longrightarrow Me_3C-Cl$$

Scheme 8.4: Retrosynthesis of *tert*-butyl alcohol.

This would be the logical mechanism, whereas if any other TM bonds such as C–Me are
broken, the intermediate species Me^+ and Me_2C^-OH would probably not exist (Scheme 8.5).
Another example: Diethyl 2-benzylmalonate is shown in Fig. 8.1.
Of the two possible disconnections, breaking bonds by (a) or (b), the best disconnection is (b), since it leads to an anion and a cation, both stabilized. Therefore, the
correct disconnection would be as in Scheme 8.6.

Scheme 8.5: Illogical disconnection of *tert*-butyl alcohol.

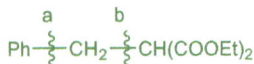

Fig. 8.1: Two possible disconnections of diethyl 2-benzylmalonate.

Scheme 8.6: Correct disconnection of diethyl 2-benzylmalonate.

The preparation of the molecule would therefore be through a malonic ester synthesis (Scheme 8.7).

Scheme 8.7: Synthesis of diethyl 2-benzylmalonate.

Another type of reaction in which it is very easy to deduce the disconnection are the Diels–Alder reactions: for example, the reaction between 1,3-butadiene and acrolein would be as shown in Scheme 8.8.

Scheme 8.8: Diels–Alder reaction.

We now formulate the disconnection of the product (Scheme 8.9).

Scheme 8.9: Disconnection of the Diels–Alder product.

The double bond in the six-membered ring tells us where to start the disconnection. The Diels–Alder reaction is a *syn* addition, both with respect to diene and dienophile (Scheme 8.10).

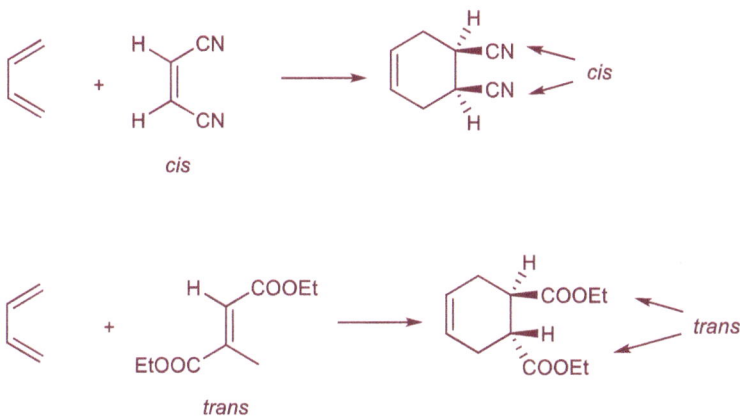

Scheme 8.10: *syn* addition of the Diels-Alder product.

8.5 *endo* selectivity

When a diene reacts with a dienophile, two isomeric Diels–Alder products called *endo* and *exo* adducts can be formed. When the electron-withdrawing group of the dienophile is next to the new double bond formed, the compound is called *endo*. When the electron-withdrawing group of the dienophile is far from the new double bond, the compound is called *exo* (Scheme 8.11).

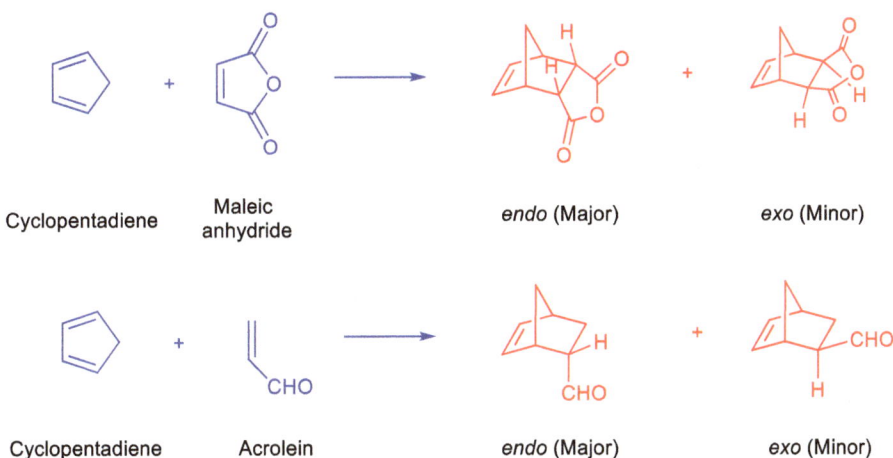

| Cyclopentadiene | Maleic anhydride | *endo* (Major) | *exo* (Minor) |

| Cyclopentadiene | Acrolein | *endo* (Major) | *exo* (Minor) |

Scheme 8.11: *Endo* and *exo* adducts of the Diels–Alder reaction.

We will use these concepts in the synthesis of buprenorphine.

8.6 Synthesis of various barbiturates

8.6.1 Pentobarbital

In the case where we have branched substituents, such as pentobarbital used as an antiepileptic, hypnotic, and sedative drug (see Chapter 5 of Volume 2 of this series), alkylation of the diethyl malonate in a basic medium with 2-bromopentane cannot be carried out, because being a secondary halide the elimination reaction predominates. In these cases, Knoevenagel reaction and subsequent reduction of the double bond yield the desired product. The introduction of the ethyl substituent is carried out by the classical method from ethyl 2-(2-pentyl)malonate (Scheme 8.12).

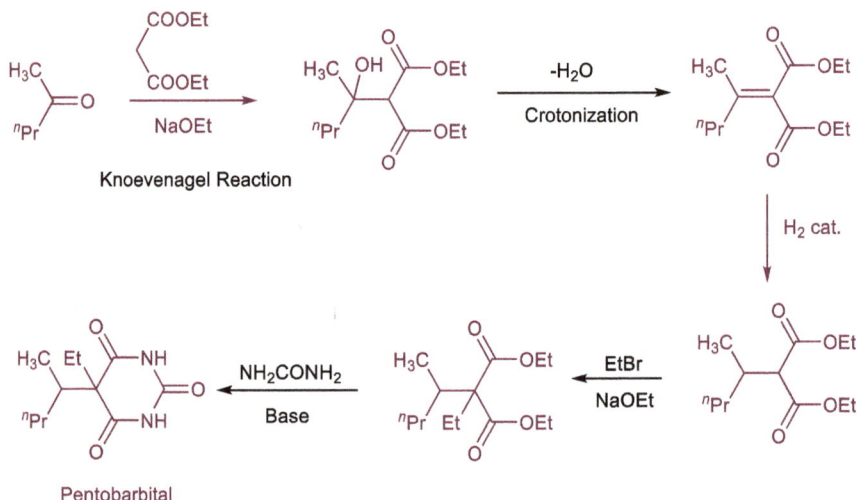

Scheme 8.12: Synthesis of pentobarbital.

8.6.2 Phenobarbital

An interesting case is the preparation of phenobarbital, the first synthesized barbiturate (see Chapter 5 of Volume 2 of this series). The preparation of the diethyl phenylmalonate is not made by direct alkylation with phenyl halide, since these derivatives are very unreactive against S_N2. 1,3-Dicarbonyl compounds can be disconnected through the α,β-bond (Scheme 8.13), and the retrosynthetic analysis leads to the carbanion of ethyl phenylacetate and diethyl carbonate.

For 1,3-dicarbonyl compounds:

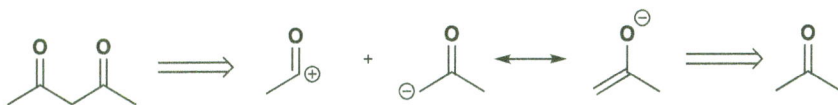

For the molecule:

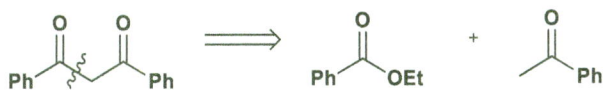

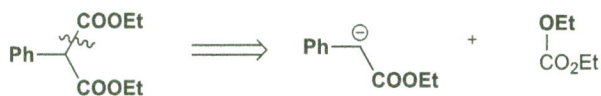

Compound CO(OEt)$_2$ is diethyl carbonate, readily available. Instead:

Scheme 8.13: Retrosynthesis of 1,3-dicarbonyl compounds.

Scheme 8.14 shows the synthesis of phenobarbital.

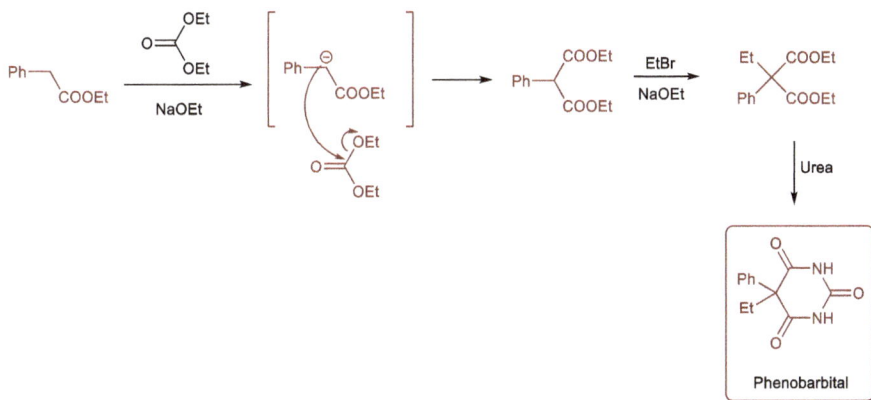

Scheme 8.14: Synthesis of phenobarbital.

8.6.3 Hexobarbital

Hexobarbital is a barbiturate derivative having hypnotic and sedative effects (see Chapter 5 of Volume 2 of this series). The introduction of the 1-cyclohexenyl substituent at the α-position of the malonic ester is affected by a Knoevenagel condensation from cyclohexanone. However, the conjugated diester is isomerized in the basic medium to the nonconjugated product, which in this case is more stable, because it is not as destabilized by steric interactions as the conjugated tautomer (Scheme 8.15).

For 1,2-unsaturated substituents, as in the case of hexobarbital:

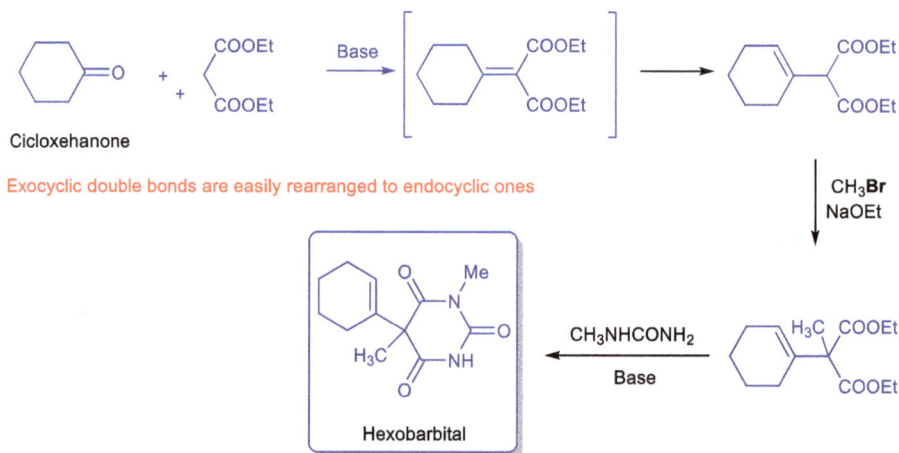

Exocyclic double bonds are easily rearranged to endocyclic ones

Scheme 8.15: Synthesis of hexobarbital.

8.7 Buprenorphine

Buprenorphine (see Chapter 6 of Volume 2 of this series) may be used as an alternative to methadone in the treatment of withdrawal symptoms, and is the result of the formation of an additional six-membered cycle between positions 6 and 14 of the opiate skeleton. From a retrosynthetic point of view, this additional cycle with a double bond and with the acetyl moiety can be prepared by means of the Diels–Alder reaction of the opioid scaffold with the methyl vinyl ketone. Finally, the reaction of resulting ketone with Grignard reagents can lead to the so-called Bentley derivatives.

Scheme 8.16 shows the synthesis of buprenorphine from oripavine via its conversion to the *endo* Diels–Alder adduct **8.1**, the saturated derivative **8.2**, the *O*-carbonate **8.3**, and the *tert*-butyl carbinol **8.4**. The palladium-catalyzed *N*-demethylation of **8.4** in the presence of cyclopropylmethylcarboxylic acid anhydride gives rise to the acyl amide **8.5**, whose treatment with Vitride® furnish buprenorphine (**8.6**).

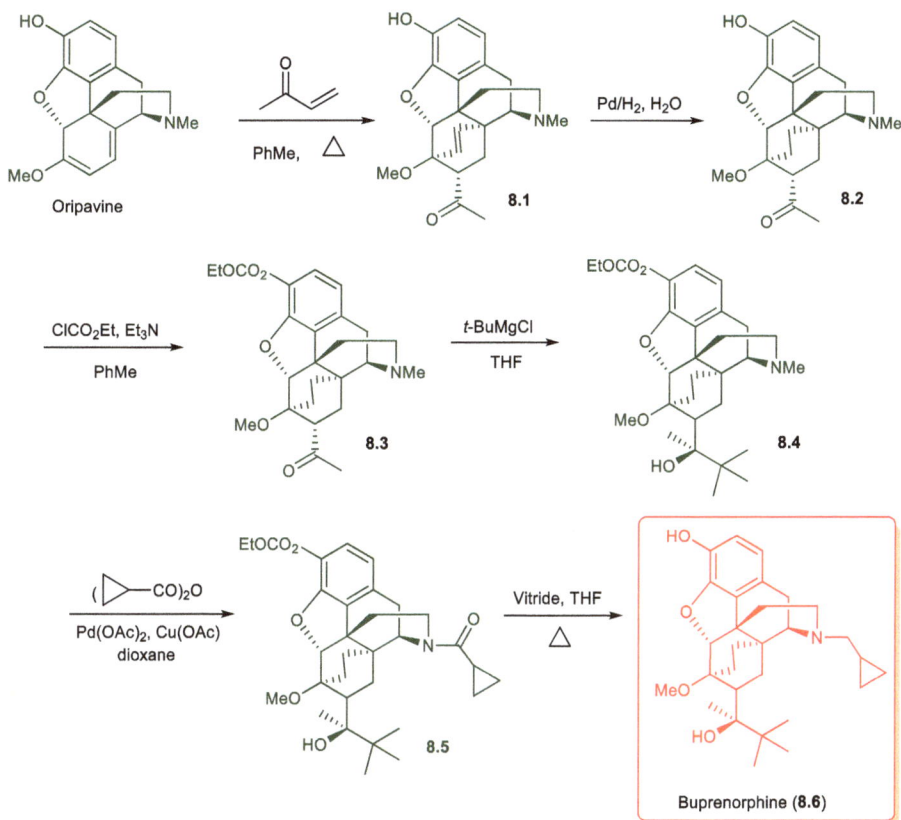

Scheme 8.16: Synthesis of buprenorphine.

8.8 Celecoxib

Celecoxib is a nonsteroidal anti-inflammatory drug (NSAID) used to treat mild-to-moderate pain and help relieve symptoms of arthritis, such as inflammation, swelling, stiffness, and joint pain (see Chapter 9 of Volume 2 of this series). The C-heteroatom bond, generally O, N, or S, in a hydrocarbon chain or cycle is usually the most suitable point for dissociation into the heterocyclic compounds. We will raise the problem of carrying out the retrosynthesis of a COX-2 inhibitor, such as celecoxib (Scheme 8.17).

Scheme 8.18 describes one of the syntheses of celecoxib. Dione **8.7** is prepared by Claisen condensation between 4-methylacetophenone and ethyl trifluoroacetate in the presence of NaOMe in methanol at reflux. The formation of the diarylpyrazole from the condensation between β-diketone **8.7** and 4-sulfonamidophenylhydrazine hydrochloride then leads to celecoxib. The corresponding chlorophenylpyrazolyl analog **8.8** is potent ($IC_{50} = 0.01$ μM vs COX-2), selective ($IC_{50} = 17.8$ μM vs COX-1), and effective. However, its

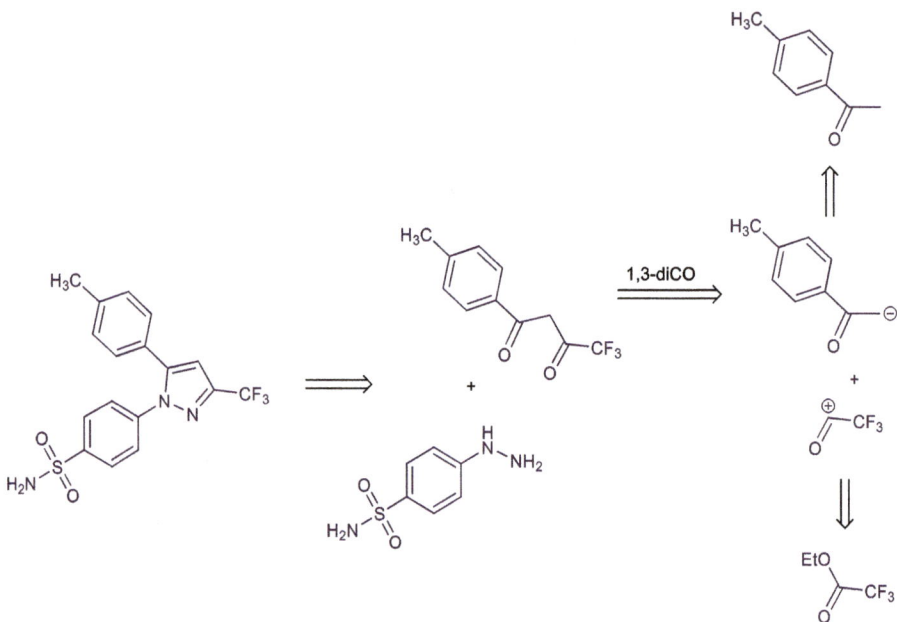

Scheme 8.17: Retrosynthesis of celecoxib.

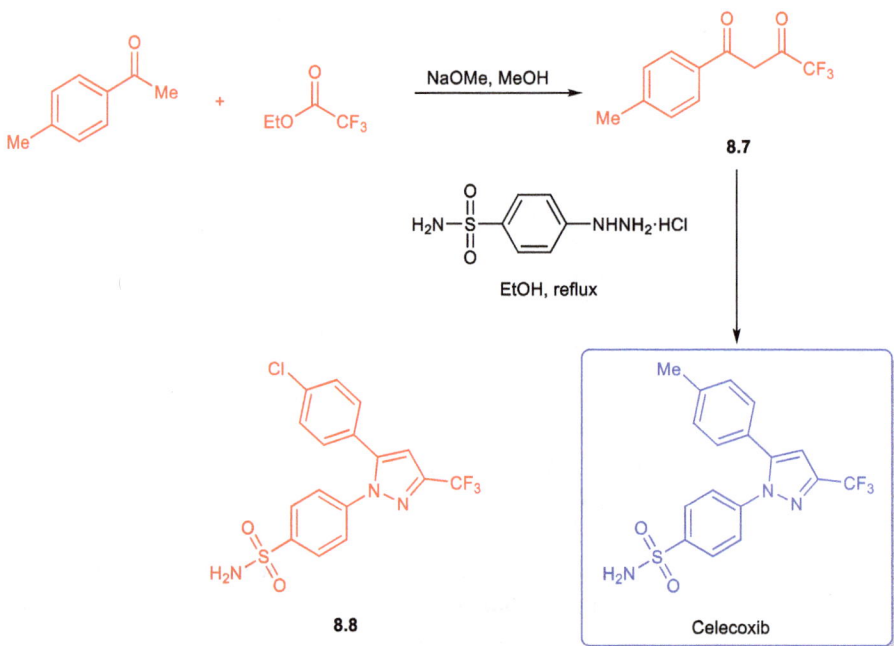

Scheme 8.18: Synthesis of celecoxib.

plasma half-life is too long. Substitution of the chlorine atom by the methyl group accelerates the metabolism to give benzoic acid via an in vivo oxidation process.

8.9 Rofecoxib

Rofecoxib (Vioxx®) is an NSAID prescribed to decrease pain and inflammation in rheumatoid arthritis (see Chapter 9 of Volume 2 of this series). In the original Merck patent published in 1995, rofecoxib is synthesized in three steps from 4-methylthioacetophenone **8.9**, prepared in turn from the Friedel–Crafts acylation of thioanisole. As shown in Scheme 8.19, the use of an excess of magnesium monoperoxyphthalate hexahydrate (a cheap, safe, and commercially available substitute for *m*-chloroperbenzoic acid) results in sulfone **8.10**, which is brominated to give compound **8.11**. This phenacyl bromide is cyclized with phenylacetic acid under the influence of 1,8-diazabicyclo[5.4.0]undec-7-ene to yield rofecoxib. The ketoester, which is formed from esterification, undergoes an intramolecular Claisen condensation to yield the furanone ring of rofecoxib.

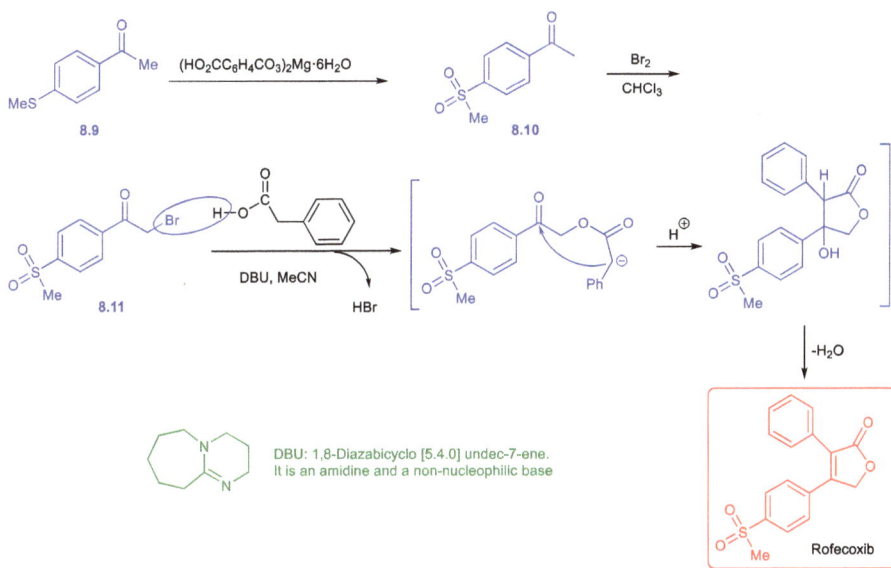

Scheme 8.19: Synthesis of rofecoxib.

8.10 Fentanyl

Fentanyl is an opiate synthetic narcotic agonist used in medicine for its analgesic and anesthetic actions (see Chapter 6 of Volume 2 of this series). Scheme 8.20 shows its retrosynthetic process.

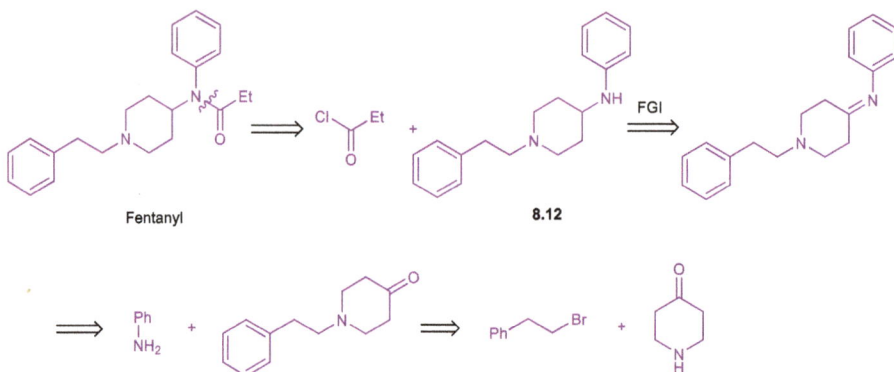

Scheme 8.20: Retrosynthesis of fentanyl.

The fentanyl synthesis is performed in three steps, starting with 4-piperidone hydrochloride. It reacts with phenethyl bromide, giving rise to *N*-phenyl-4-piperidone, which is subsequently subjected to a reductive amination to give **8.12**. Finally, **8.12** reacts with propionic chloride to yield fentanyl (Scheme 8.21).

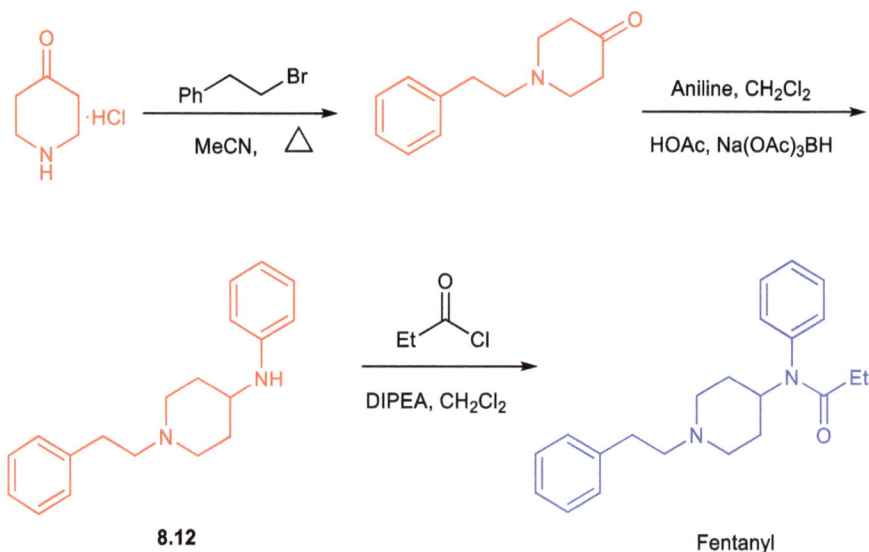

Scheme 8.21: Synthesis of fentanyl.

8.11 Electrophilic substitution on pyrroles

Thiophene, pyrrole, and furan can be considered as aromatic, to a greater or lesser degree, according to their resonance energies (Scheme 8.22). As it has been discussed in Chapter 4, all of them present an excess of π-electrons, since they have six electrons distributed in the five atoms of the ring. Thus, they are π-excessive heterocyclic compounds. Therefore, due to this, their chemistry presents certain parallels to benzene aromatic compounds, and nucleophilic compounds such as aniline or phenol.

The aromaticity of these heterocycles depends on the two electrons that the heteroatom contributes to the π-system.

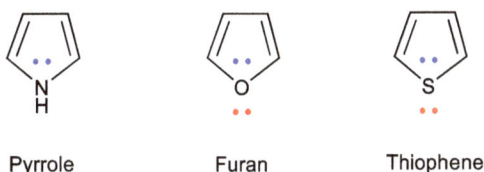

| Pyrrole | Furan | Thiophene |

Blue electronic pairs participate in the aromaticity
Red Electronic pairs do not participate in the aromaticity

1.- Resonance energy:

Furan	21 Kcal.mol^{-1}
Pyrrole	24 Kcal.mol^{-1}
Thiophene	31 Kcal.mol^{-1}
Benzene	36 Kcal.mol^{-1}

2.- Therefore, the aromatic character of these three heterocycles is as follows:

Thiophene > Pyrrole > Furan

3.- This order is related to the electronegativity values (Pauling scale of electronegativy):

Element	O	N	S
Electronegativity	3.44	3.04	2.58

4.- As O is more electronegative than N and S, it provides the two electrons needed for the aromaticity less easily, and consequently it is less aromatic than pyrrole and thiophene

Scheme 8.22: Structures of thiophene, pyrrole, and furan.

Electrophilic aromatic substitution normally occurs at position C2, which is next to the heteroatom, through the more stable intermediate (Scheme 8.23). When position 2 is occupied, the substitution takes place at position 5.

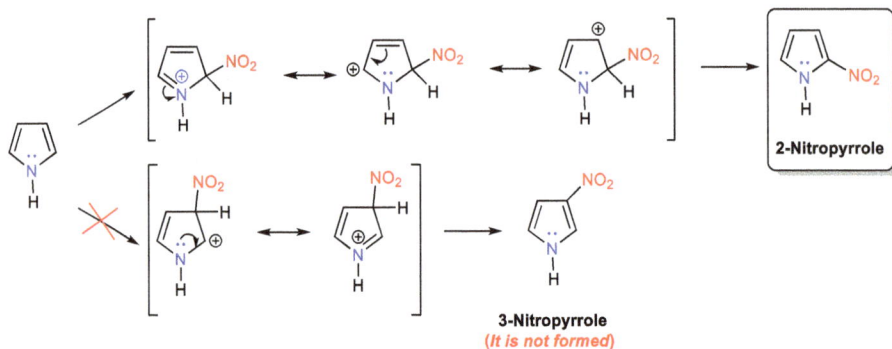

Scheme 8.23: Substitution position on five-membered aromatic compounds.

The three five-membered heteroaromatic compounds react all too easily with electrophiles and are unstable in acid whether protic or Lewis. Therefore, we must find reactions that can be used in neutral or weakly acidic solutions. The synthesis of the NSAID tolmetin illustrates these two features. The disconnection of the ketone would lead naturally to an $AlCl_3$-catalyzed Friedel–Crafts reaction between the acid chloride and the pyrrole derivative (Scheme 8.24).

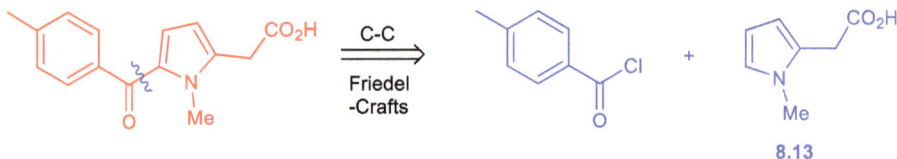

Scheme 8.24: Disconnection of tolmetin.

But this mixture would decompose the pyrrole **8.13**. The Vilsmeier acylation has to be used, replacing the acid chloride by the tertiary amide and $AlCl_3$ by $POCl_3$. The amide is very unreactive but combines with $POCl_3$ to give a reactive species that does attack pyrrole **8.15** at position 5 to give, after re-aromatization and hydrolysis, the ketone **8.16** (Scheme 8.25).

    The pyrrole with the alkyl side chain must be prepared for this acylation reaction. Friedel–Crafts alkylation is not an option, but instead, Mannich reaction can be carried out on pyrroles: Formaldehyde and an amine combine to give another iminium salt that reacts with N-methyl pyrrole to give, after re-aromatization, the substituted pyrrole **8.17** (Scheme 8.26).

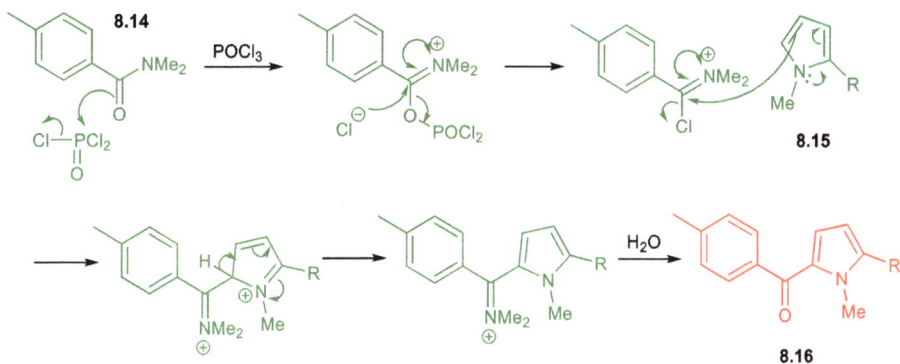

Scheme 8.25: Application of the Vilsmeier acylation on *N*-methylpyrrole.

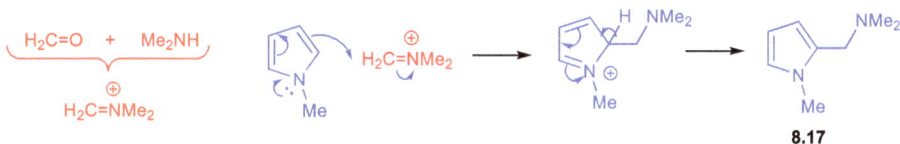

Scheme 8.26: Synthesis of *N,N*-dimethyl-1-(1-methyl-1*H*-pyrrole-2-yl)methanamine.

The tertiary amino group in Mannich products is often replaced by other functionalities: Here methylation and substitution are used to make nitrile **8.18**, and this compound was used in the acylation sequence. Finally, hydrolysis of nitrile **8.18** gave tolmetin (Scheme 8.27).

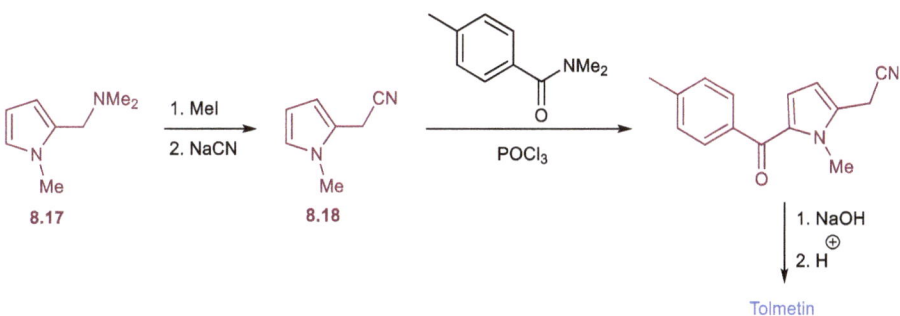

Scheme 8.27: Conversion between *N,N*-dimethyl-1-(1-methyl-1*H*-pyrrole-2-yl)methanamine and tolmetin.

8.12 Exercises

1. Propose the retrosynthesis and synthesis of the following drugs:

(a) (*RS*)-fluoxetine (Prozac®)

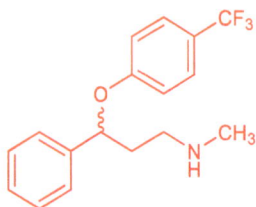

(b) Nifedipine

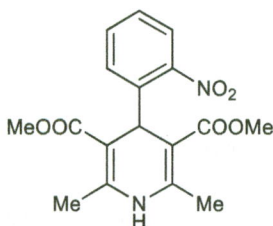

(c) Trimethoprim

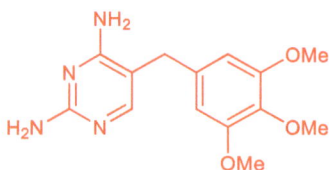

(d) Methotrexate

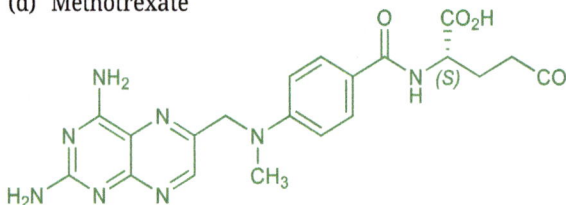

(e) Sumatriptan

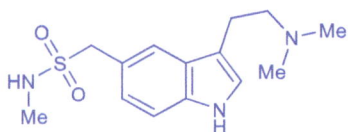

9 Solutions to the exercises

9.1 Exercises to Chapter 2

A. Name the following drugs:
    (1) 4-Amino-*N*-[2-(diethylamino)ethyl]benzamide
    (2) 4-Amino-5-chloro-*N*-(2-diethylaminoethyl)-2-methoxybenzamide
    (3) 2-Amino-3,5-dibromo-*N*-cyclohexyl-*N*-methylbenzylamine
    (4) 7-(Cyclohexylcarbonyl)-2-naphthoic acid
    (5) *N*-(5-Nitro-2-propoxyphenyl)acetamide
    (6) Ethyl 2-phenoxy-2-methylpropanoate
    (7) 6-(Dimethylamino)-5-methyl-4,4-diphenylhexane-3-one
    (8) 2-(4-Chlorophenoxy)-2-methylpropanoic acid
    (9) (2-Amino-3-benzoylphenyl)acetic acid
    (10) 1-(2,6-Dimethylphenoxy)propan-2-amine
    (11) 6-(Phenoxypropanamido)-3,3-dimethyl-7-oxo-4-thia-1-azabicyclo[3.2.0]heptane-2-carboxylic acid
    (12) 4,7-Dimethyl-7-azabicyclo[2.2.1]heptan-2-yl propionate
    (13) 7-[2-(Thiophene-2-yl)acetamido]-3-[(carbamoyl)methyl]-8-oxo-5-thia-1-azabicyclo[4.2.0]oct-2-ene-2-carboxylic acid
    (14) 3-[(2-Aminoethyl)thio]-6-(1-hydroxyethyl)-7-oxo-1-azabicyclo[3.2.0]hept-2-en-2-carboxylic acid
    (15) 2-Methoxycarbonyl-8-methyl-8-azabicyclo[3.2.1]octane-3-yl benzoate
    (16) Propyl 8-methyl-8-azabicyclo[3.2.1]octane-3-yl-pentanoate
    (17) 8-[4-(4-Fluorophenyl)-4-oxobutyl]-1-phenyl-1,3,8-triazaspiro[4.5]decan-4-one
    (18) 5-[(2-Methoxyphenoxy)methyl]oxazolidine-2-one
    (19) 1-(*p*-Chlorobenzyl)-2-[(1-pyrrolidinyl)methyl]-1*H*-benzimidazole
    (20) 8-Chloro-11-(4-methylpiperazin-1-yl)-10,11-dihydro-5*H*-dibenzo[*b,e*]-1,4-diazepine
    (21) 5-(4-Chlorophenyl)-3,5-dihydro-2*H*-imidazo[2,1-*a*]isoindole-5-ol
    (22) 11-[3-(Dimethylamino)propylidene]-6,11-dihydrodibenzo[*b,e*]thiepine
    (23) 2-Bromo-4-(2-chlorophenyl)-9-methyl-6*H*-thieno[3,2*f*][1,2,4]triazolo[4,3-*a*]-1,4-diazepine
    (24) 7-(α-Amino-α-(3-hydroxyphenyl)acetamido)-8-oxo-3-(1-propene-1-yl)-5-thia-1-azabicyclo[4.2.0]oct-2-ene-2-carboxylic acid
    (25) 2-(6-Chloro-9*H*-carbazol-2-yl)propionic acid
    (26) 8-Chloro-10-(2-dimethylaminoethoxy)dibenzo[*b,f*]thiepine
    (27) 10-(2-Dimethylaminopropyl)-10*H*-pyrido[3,2-*b*][1,4]benzothiazine or 1-(10*H*-benzo[*b*]pyrido[2,3-*e*]-1,4-thiazine-10-yl)-*N,N*-dimethylpropan-2-amine
    (28) 4-Amino-2-[4-(furan-2-carbonyl)-1-piperazinyl]-6,7-dimethoxyquinazoline or {4-[4-amino-6,6-dimethoxyquinazoline-2-yl]piperazine-1-yl}(furan-2-yl)methanone

https://doi.org/10.1515/9783111316901-009

(29) 6-Benzoyl-3-hydrazino-5,6,7,8-tetrahydropyrido[4,3-*c*]pyridazine or (3-hydrazine
   -yl-7,7-dihydropyrido[4,3-*c*]pyridazine-6(5*H*)-yl)phenylmethanone

(30) 2-[3-(4-Chlorophenyl)-1-phenyl-1*H*-pyrazole-4-yl]acetic acid

(31) *N*¹-(6-Methoxy-2-methylpyrimidine-1-methyl)sulfanilamide

(32) *N*¹-(5-Methyl-1,3,4-thiadiazole-2-yl)sulfanilamide

(33) Ethyl methyl 4-(2,3-dichlorophenyl)-2,6-dimethyl-1,4-dihydropyridine-3,5-
   dicarboxylate

(34) 7-Chloro-*N*-(5-diethylamino)pentan-2-yl)quinazoline-4-amine

(35) 10-Chloro-11*b*-(2-chlorophenyl)-7-methyl-3,5,7,11*b*-tetrahydro-2*H*-oxazolo[3,2-
   d]-1,4-benzodiazepine-6-one

(36) 3-(Acetoxymethyl)-7-[2-(2-thienyl)acetamido]-8-oxo-5-thia-1-azabicyclo[4.2.0]
   oct-2-ene-2-carboxylic acid

B. Formulate the following drugs:

(1)

(2)

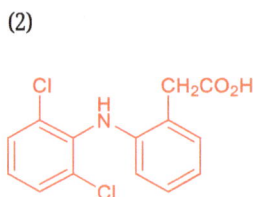

(3)

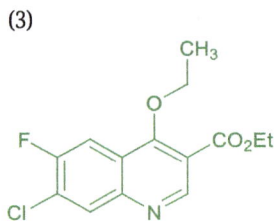

(4)

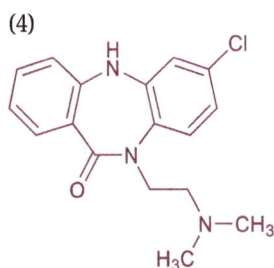

(5)

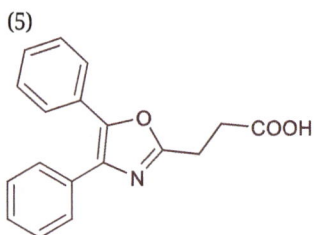

(6)

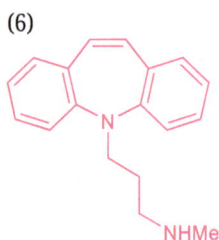

(7)

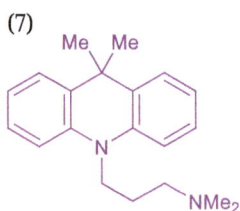

(8)

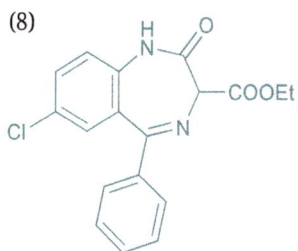

(9)

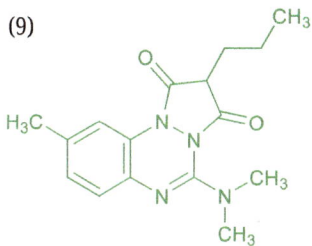

(10)

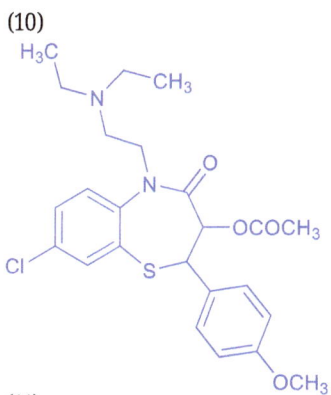

(11)

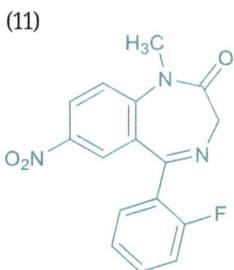

(12)

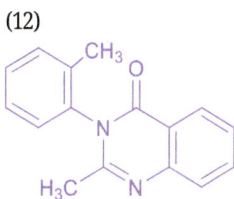

(13)

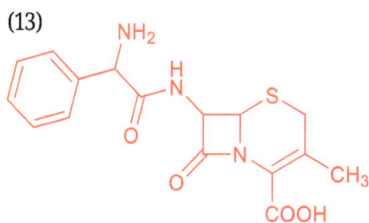

(14)

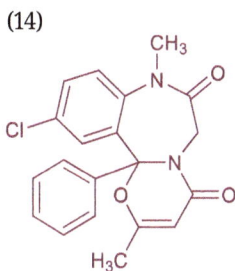

(15)

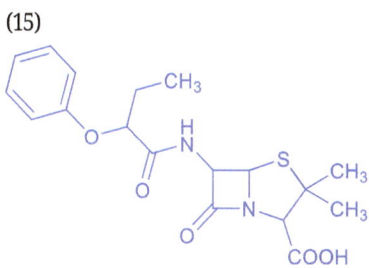

(16)

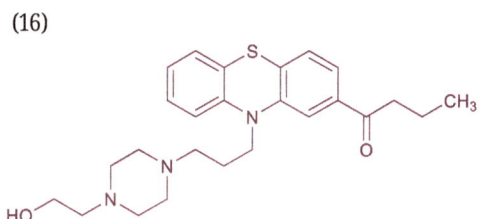

(17)

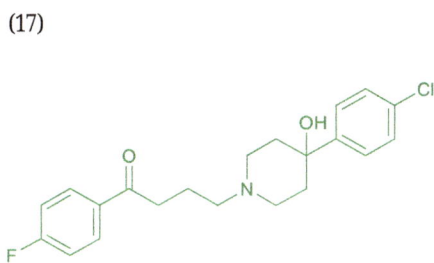

(18)

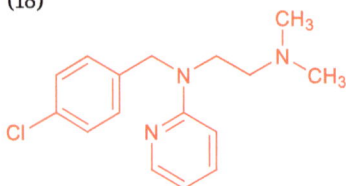

(20)

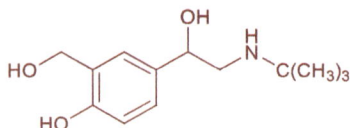

(19)

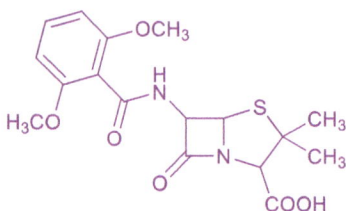

9.2 Exercises to Chapter 4

1. (a) Modulative structural modification due to a ring size variation
    (b) Modulative variation due to reorganization and dissociation of the rings
    (c) Modulative structural variation by isomerization
    (d) Modulative structural variation by formation of a rigid analog
    (e) Conjunctive replication by formation of a hybrid or a molecular combination
    (f) Structural modulative variation by bioisosterism
    (g) Disjunctive replication
    (h) Disjunctive replication
    (i) Modulative structural variation by formation of a cycle
    (j) Conjunctive replication: molecular duplication
    (k) Modulative structural variation by bioisosterism
    (l) Conjunctive replication by association of two molecules (antihypertensive + diuretic drug)
    (m) Disjunctive replication (simplification of the structure)
    (n) Modulative structural variation due to the formation of a cycle
    (o) Modulative variation by a bioisosteric change (suprophen: a drug with analgesic and an anti-inflammatory property)

2. (a) Vinylogous: procaine and its active vinylogue
    (b) Isosteres: change of the carbonyl and sulfone groups (meperidine and its isostere that is also a hypnoanalgesic drug)
    (c) Isosteres: indomethacin and oxametacin (an isosteric anti-inflammatory agent)
    (d) Homologs: procaine and its inactive counterpart
    (e) Isosteres: catechol and benzimidazole

    (f) Isosteres: 4-dimethylaminoantipyrine and 4-isopropylantipyrine, both analgesic and antipyretic agents

    (g) Homologs

    (h) Isosteres: hypoxanthine and 6-mercaptopurine

    (i) Isosteres: methaphenilene and methapyrilene

    (j) Vinylogues: pethidine (meperidine) and its vinylogue

    (k) Vinylogues: phenylbutazone and styrylbutazone

3. (a) Modification of pharmacokinetics by blocking the metabolism by the introduction of fluorine atoms into the aromatic rings: longer duration of action. The compound on the left is CGP52411, an enzymatic inhibitor that acts on the active site of the epidermal growth factor receptor, useful as an antitumor agent. This compound undergoes oxidative metabolism at the *para*-position of the aromatic ring. The introduction of F resulted in **CGP53353**. This same procedure was successfully applied in the synthesis of gefitinib, which is a drug used for certain breast, lung, and other cancers.

    (b) Transformation of an agonist into antagonist (hypoxanthine, a metabolite, and 6-mercaptopurine, an antitumor agent by antagonism)

    (c) Development of therapeutic copies: enalapril and its therapeutic copy, quinapril

    (d) Activity change (homology allowed by chance the interaction with another biological target): dietazine, an H_1 antihistaminic drug, and chlorpromazine, a neuroleptic change

    (e) Reduction of toxicity: the change of chlorine atoms of the compound **UK47265** by F atoms led to the antifungal fluconazole, lacking toxicity

    (f) Improvement of distribution so that it reaches the organs where it has to exercise its action: mechlorethamine and melphalan

    (g) Change of the spectrum of action: transformation of an agonist into an antagonist, by introduction of bulky groups (acetylcholine and propantheline)

    (h) Improvement of distribution: the compound does not cross the blood–brain barrier with the quaternization of nitrogen and acts only at the peripheral level (scopolamine and scopolamine butylbromide)

9.3 Exercises to Chapter 5

1. (a)

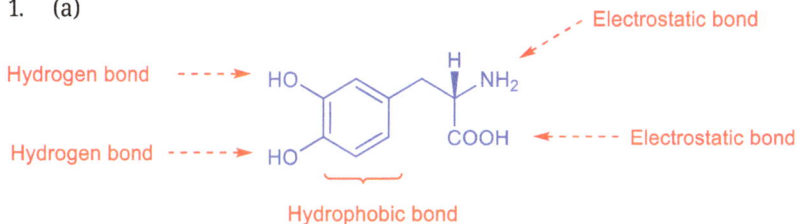

(b)

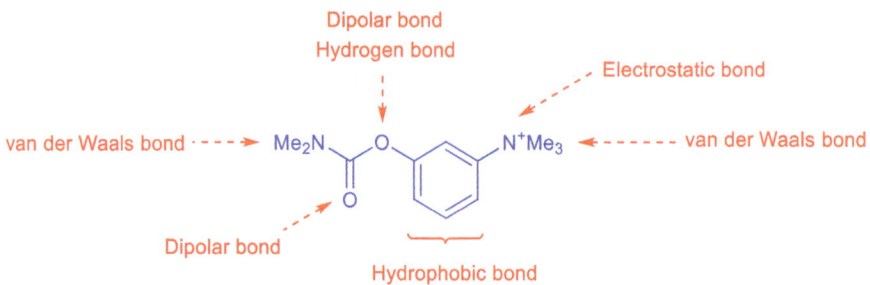

Dipolar bond
Hydrogen bond

Electrostatic bond

van der Waals bond ·····▸ Me₂N O N⁺Me₃ ◂······· van der Waals bond

Dipolar bond

Hydrophobic bond

(c)

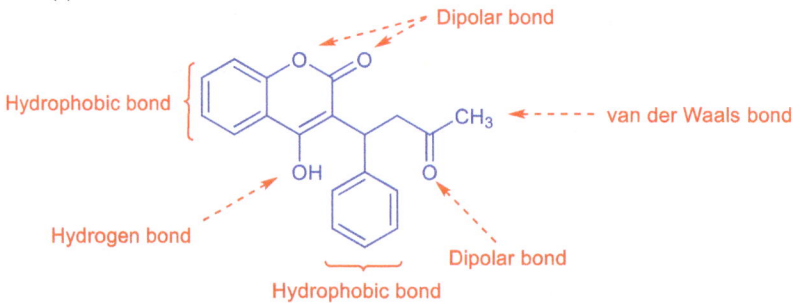

Dipolar bond

Hydrophobic bond

CH₃ ◂······ van der Waals bond

Hydrogen bond

Hydrophobic bond

Dipolar bond

(d)

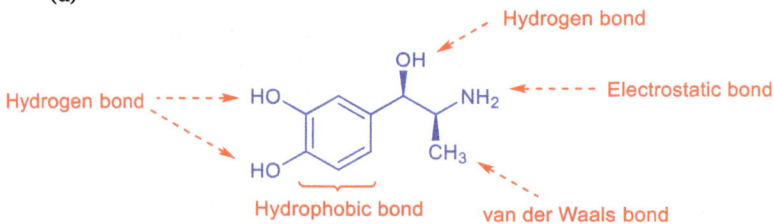

Hydrogen bond

OH

Hydrogen bond ·····▸ HO NH₂ ◂····· Electrostatic bond

HO CH₃

Hydrophobic bond van der Waals bond

(e)

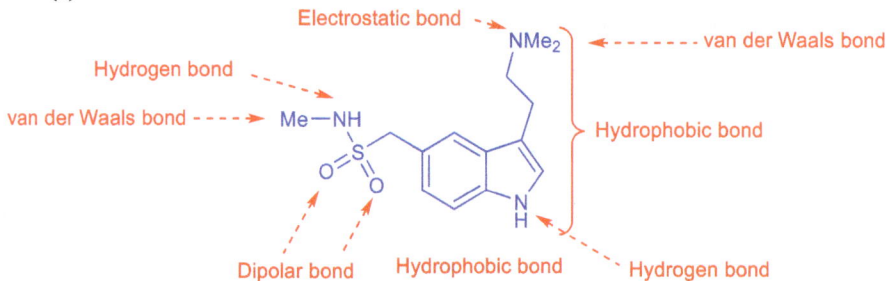

Electrostatic bond

NMe₂ ◂······· van der Waals bond

Hydrogen bond

van der Waals bond ·····▸ Me—NH

Hydrophobic bond

Dipolar bond Hydrophobic bond Hydrogen bond

(f)

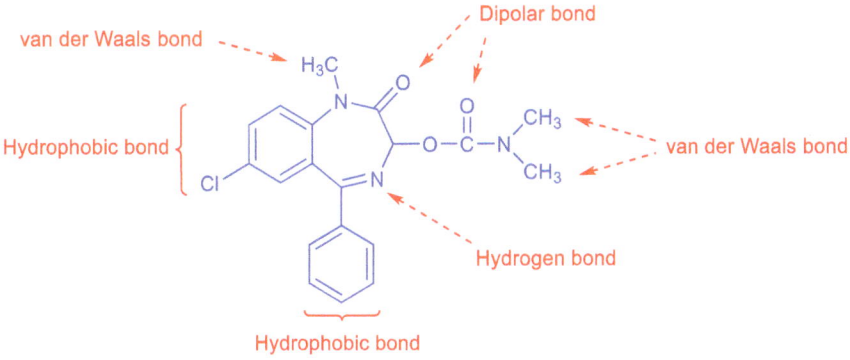

(g)

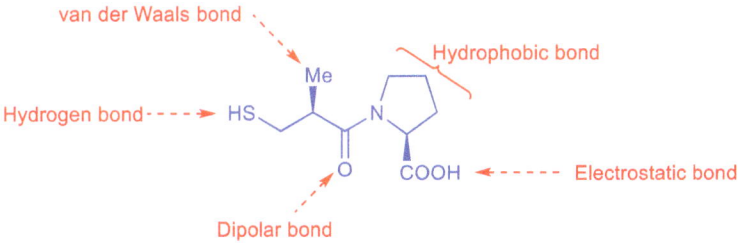

(h)

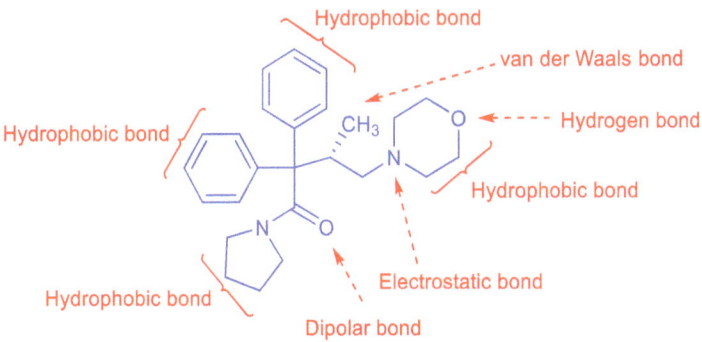

2. (a)

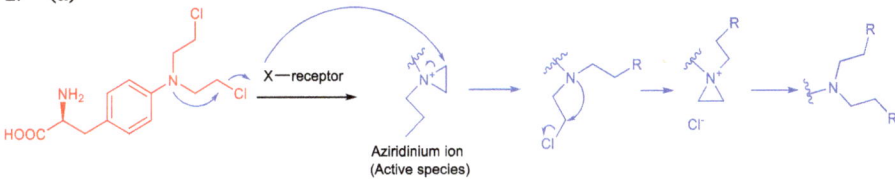

(b)

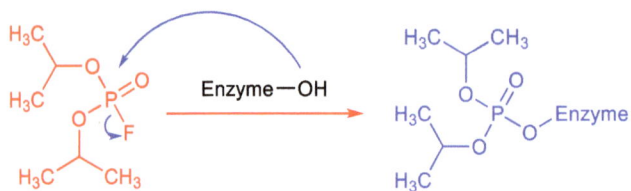

(c)

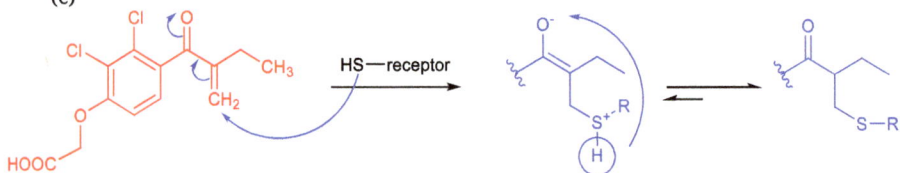

(d)

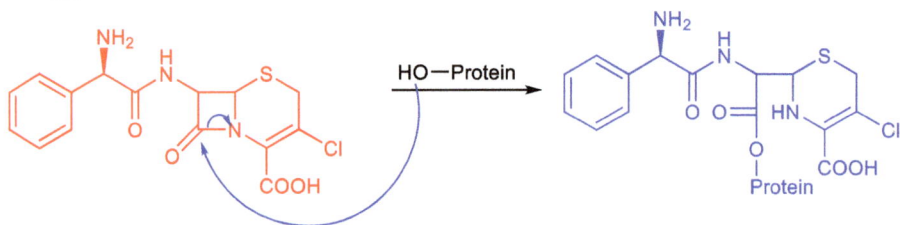

(e)

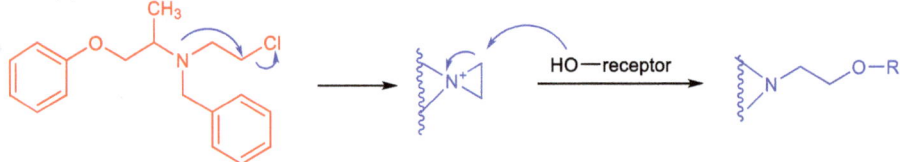

3.

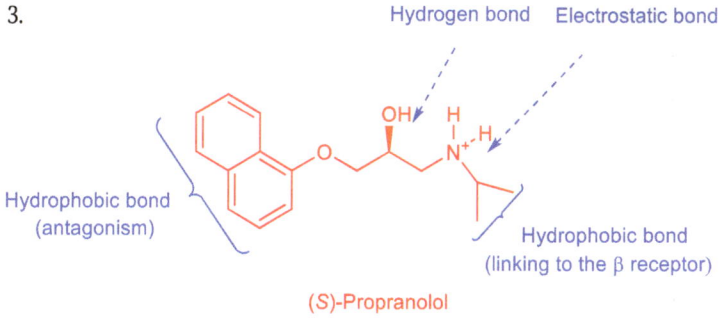

(S)-Propranolol

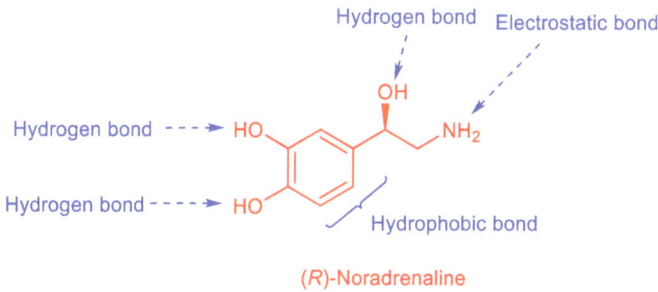

(R)-Noradrenaline

4. (a)

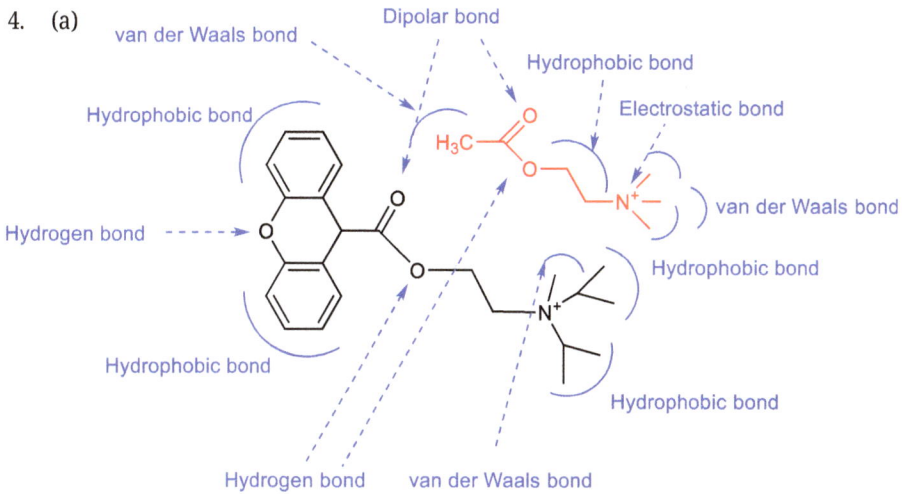

Hydrophobic interactions cause it to act as an antagonist.

(b)

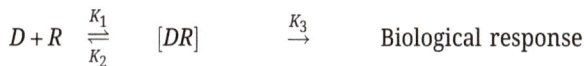

Agonist: big $K_1 \geq K_2$, K_3
Antagonist: $K_1 \geq K_2$, $K_3 = 0$

Partial agonist: small $K_1 \geq K_2, K_3$
Inactive compound: $K_1 \leq K_2$

5. (a)

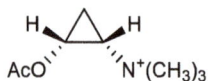 

  cis   Diastereomers   *trans*

  Nicotinic agonist   Muscarinic agonist

                                      It mimics the antiperiplanar
                                      conformation of acetylcholine

  (b)

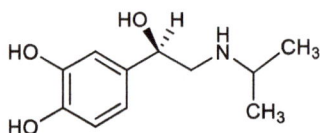 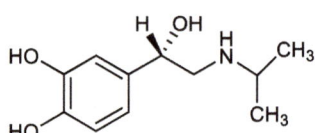

        R(-)   *S(+)*

                          Enantiomers of
  Potent β agonist   isoprenaline   Weak β agonist

  Weak α antagonist   α Antagonist

Eutomer (bronchodilator activity)

  (c)

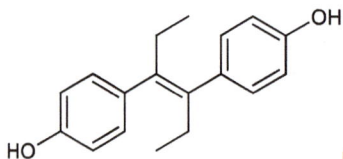 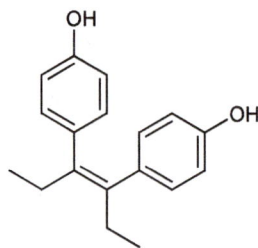

                          Diethylstilbestrol
        trans   diastereomers   *cis*

Higher estrogenic activity; it has the   Weakly active
OH groups oriented as in estradiol

(d)

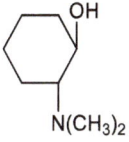

OH
N(CH₃)₂
2-Dimethylaminocyclohexanol
trans

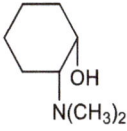

OH
N(CH₃)₂
cis

Diastereomers
Geometric and optical isomers
Enzymatic inhibidors
They have different
coupling ability to the enzyme

(e)

H_3C—S—O—...—O—S—CH_3

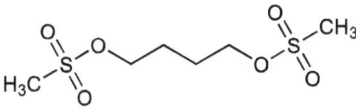

Busulfan (antineoplastic agent)

It acts by alkylation of nucleophilic
groups of the biological target

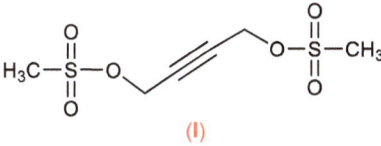

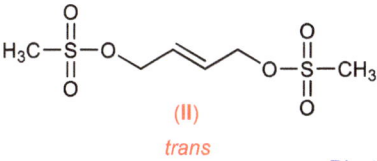

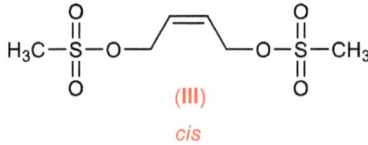

H_3C—S—O—...—O—S—CH_3
(I)

H_3C—S—O—...—O—S—CH_3
(II)
trans

H_3C—S—O—...—O—S—CH_3
(III)
cis

Diastereomers
Rigid busulfan analogs

NH₂ ///////

MsO———————OMs

N ///////

MsO———OMs
NH₂ ///////

N ///////

Only the *cis*-isomer has potency similar to that of the flexible drug. It follows that the antineoplastic activity is related to the ability to form a pyrrolidine derivative by bis-1,4-alkylation of some amino groups suitably disposed in the biological target.

(f)

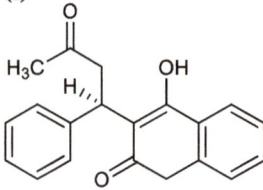
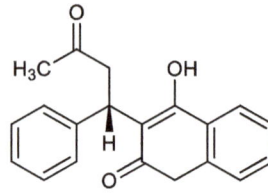

(*R*)-Warfarin   Enantiomers   (*S*)-Warfarin

Distomer   Eutomer
(Anticoagulant agent)

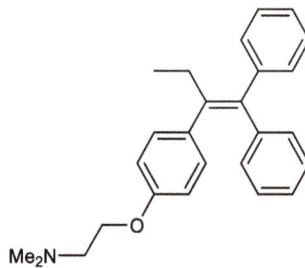
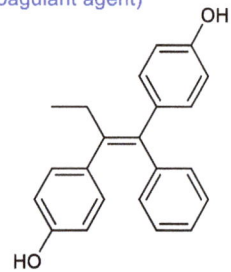

Tamoxifen   Tamoxifen metabolite

6.

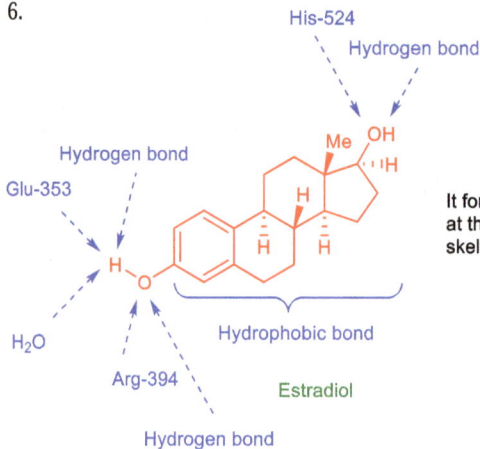

It forms hydrogen bonds with 3 amino acids at the receptor binding site and the lipophilic skeleton forms hydrophobic bonds with other regions

The hydrophobic pocket is quite spacious, except in the area where the phenol binds: this is a narrow zone, which only accepts a flat ring, and due to this the phenolic ring determines the orientation of the rest of the molecule.

Tamoxifen metabolite would bind to the estrogen receptor in the same way as estradiol, while tamoxifen, with its dimethylaminoethyl group, would not fit into the receptor hole. In addition, it could not form all the hydrogen bonds needed for the bond.

9.4 Exercises to Chapter 6

1. (a) *m*-Xylene (3.20)

$$\pi_{Me\,(aromatic)} = 0.56$$

$$\log P_{mesitylene} = \log P_{benzene} + 2\pi_{Me} = 2.13 + (2 \times 0.56) = 3.25$$

(b) Mesitylene (1,3,5-trimethylbenzene) (3.43)

$$\log P_{mesitylene} = \log P_{benzene} + 3\pi_{Me} = 2.13 + 1.68 = 3.81$$

(c) Hexamethylbenzene (2.33)

$$\log P_{hexamethylbenzene} = \log P_{benzene} + 6\pi_{Me} = 2.13 + 3.36 = 5.49$$

The fact that the experimental value is lower than the calculated one may be due to the large number of substituents in the benzene ring and to its electron-donating effect that causes this partition coefficient to fall.

(d) 1,3-Dinitrobenzene (1.49)

$$\pi_{NO_2\,(aromatic)} = -0.28$$

$$\log P_{nitrobenzene} = \log P_{benzene} + 2\pi_{NO_2} = 2.13 - 0.56 = 1.57$$

(e) 2,4-Dihydroxybenzoic acid (1.44)

$$\pi_{COOH\,(aromatic)} = -0.32; \quad \pi_{OH\,(aromatic)} = -0.67$$

$$\log P_{2,4\,-\,dihydroxybenzoic\,acid} = \log P_{benzene} + \pi_{COOH} + 2\pi_{OH}$$

$$= 2.13 - 0.32 - 1.34 = 0.47$$

In this case, the difference in values is due to the presence of an intramolecular bond that increases the lipophilicity of the molecule with respect to the calculated value:

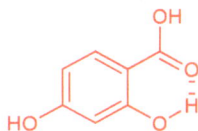

2. The negative value of the coefficient of π indicates that the activity is favored by hydrophilic substituents ($\pi < 0$); since σ has a positive coefficient, the electron-withdrawing substituents favor potency. Thus, both s polar and an electron-withdrawing group will yield a compound of optimal activity, whereas an electron-releasing group will yield a weak methicillin derivative; for example:

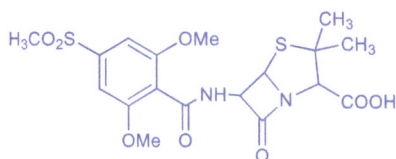

The p-SO$_2$CH$_3$ group is polar (π = -0.50) and an electron-withdrawing (σ = 0.72). If we replace it in the QSAR equation, we will obtain log (1/C) = 3.14

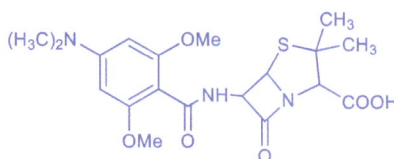

The p-N(CH$_3$)$_2$ group is slightly lipophilic (π = 0.18) and an electron-donating group (σ = -0.83). If we replace it in the QSAR equation, we will obtain log (1/C) = 0.30

The foregoing values allow the prediction that the activity of p-methylsulfonyl-methicillin would be some 700 times higher than expected for p-dimethylamino-methicillin. However, these predictions must be verified by synthesis and assay of the compounds.

3. (a) Those substituents that have a high partition coefficient (log P) and that are electron-withdrawing groups.

   (b) An electron-withdrawing substituent X will be needed to facilitate the nucleophilic attack of a thiol group (–SH):
   Mechanism of nucleophilic attack reaction:

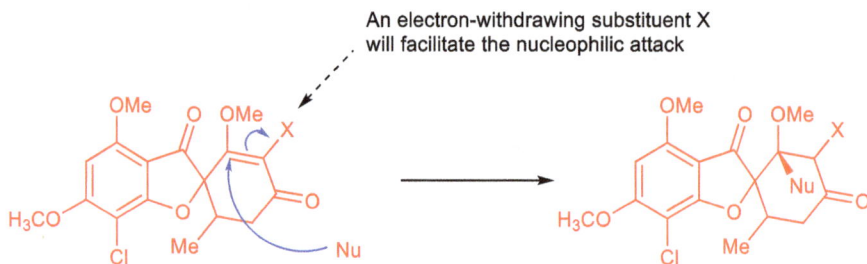

An electron-withdrawing substituent X will facilitate the nucleophilic attack

4. (a) The first equation indicates that adrenergic blocking activity depends on lipophilicity (π) and on electronic effects (σ). The second equation shows that hepatic toxicity (as well as activity) increases with the presence of substituents in the aromatic ring that increase lipophilicity (π positive) and that are electron-releasing groups (σ negative), such as –NMe$_2$.

   (b) Mechanism of alkylation:

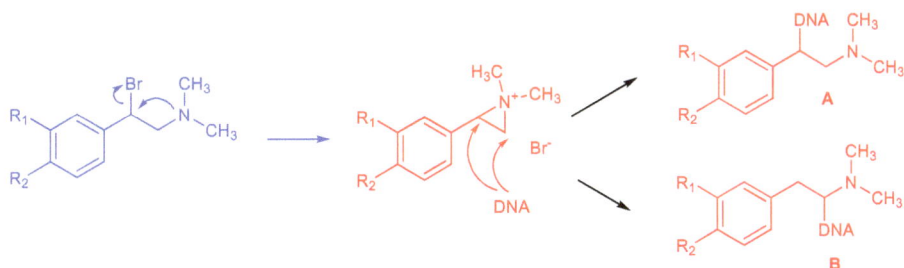

(c) Many reactions create positive charges that can be stabilized by delocalization via resonance with the substituent. Therefore, a new substituent effect scale was produced for groups that stabilized positive charges via resonance (σ^+), such as in the intermediate that gives rise to the alkylated DNA compounds **A** and **B**. The σ^+ scale is based upon the heterolysis (S_N1) reaction of *para*-substituted cumyl chlorides (phenyldimethyl chloromethanes). The rationale for using parameters σ^-, σ^+, and σ^0 is discussed in Chapter 1 of the Hansch and Leo's book, which is referenced in the fundamental bibliography in the Prologue of this volume:

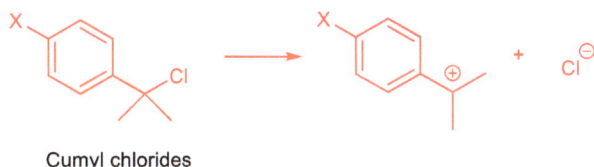

Cumyl chlorides

5. (a) The inhibitory activity depends on the distribution coefficient as this is important in the transportation of the drug at the physiological level.

   (b) To calculate the best value of π, the derivative of the equation with respect to π has to be calculated:

   $$\frac{d\left(\log\frac{1}{C}\right)}{d\pi} = 0.34 - 2(0.183\ \pi) = 0$$

   $$\pi = \frac{0.304}{0.366} = 0.831$$

   Of the proposed substituents, the closest to this value is Br (0.86). The Br substituent would be therefore the best group to improve the in *vivo* inhibitory activity.

6. (a) The second equation indicates that the activity is favored by electron-withdrawing substituents (σ positive values).

   According to the first equation, the lower the absolute value of pK_a, the higher the activity, i.e. the activity will be favored by substituents that increase the acidity of the substituents.

An electron-withdrawing R substituent in the structure of the sulfonamide moiety will stabilize the anion that would be formed by the abstraction of a proton in structure **6.3**. Thus, both equations are related to the same parameter: the acidity of the sulfonamide.

(b) The fact that the activity increments on increasing the acidity is indicative that the compound will act in its ionized form.

(c) Electron-withdrawing substituents, such as halogens (–Br, –Cl), –NO$_2$, –CF$_3$, would improve the activity.

(d) The parabolic dependence of the activity on pK_a is explained by the influence of the ionization on the penetration through the bacterial membranes. Thus, a fully ionized substituent would be very hydrophilic and would not go through the membrane. A balance between hydrophilicity and lipophilicity is required so that part of the drug will have to be in the nonionized form to cross the biological membrane.

7. It depends on hydrophobic and electronic parameters, and substituents lipophilic and electron-releasing will increase log(1/K_i):

   – for the –Cl-substituted pyrazole:

   $$\log P = 0.28 \ (\text{pyrazole}) + 0.71 \ (\text{for aromatic} - \text{Cl}) = 0.99$$

   – for the –Cl-substituted pyrazole:

   $$\log (1/K_i) = 1.22 \times 0.99 + 1.80 \times 0.37 + 4.87 = 6.75$$

8. (a) It depends on hydrophobic, electronic, and steric parameters.

   (b) The activity will be favored by lipophilic substituents ($\pi > 0$), electron-withdrawing ($\sigma > 0$), and not bulky substituents ($E_s < 0$).

   (c) Examples: –F and –CF$_3$.

9. Biological activity correlates with hydrophobic and electronic parameters. It is favored by the presence of substituents that are both lipophilic (apolar, $\pi > 0$) and electron-withdrawing groups ($\sigma > 0$), such as –Br and –SEt$_2$.

10. (a) It depends on hydrophobic, electronic, and steric parameters.

    (b) The activity will be favored by the presence of lipophilic substituents ($\pi > 0$), electron-withdrawing groups ($\sigma > 0$) and not bulky surrogates ($E_s < 0$). Examples: –CH$_2$F and –SCH$_3$.

11. Lipophilic ($\pi > 0$) and electron-withdrawing substituents ($\sigma > 0$), such as halogens (–Br, –Cl, and –I), must be introduced.

12. (a) It depends on lipophilic and steric parameters.

(b) Hydrophobic ($\pi > 0$) and quite voluminous ($E_s > 0$) substituents would be introduced. Examples: $-Pr^n$, $-Bu^n$, $-Bu^t$, and $-Ph$.

13. The activity would be improved with lipophilic substituents ($\pi > 0$) and small substituents ($E_s < 0$), such as $-CH_2Cl$, $-CH_2F$, $-CH_3$, and $-SH$.

14. (a) It depends on electronic, hydrophobic, and steric parameters. The biological activity will be favored with electron-withdrawing ($\sigma > 0$), lipophilic ($\pi > 0$), and small substituents ($E_s < 0$).
    (b) Examples: $-SCH_3$, $-F$, $-Cl$, and $-CH_2F$.

15. The QSAR equation resulting from the multiple linear regression analysis for the antimalarial activity indicates that the two substituents X and Y of the substituted aromatic rings will be affected by the same type of parameters (hydrophobic and electron), and in the same order of magnitude ($\pi > 0$ and $\sigma > 0$). Therefore, the activity will be favored by lipophilic and electron-withdrawing substituents, such as $-Cl$, $-Br$, $-I$, $-CH_2Br$, and $-CCl_3$.

9.5 Exercises to Chapter 7

1. (a)

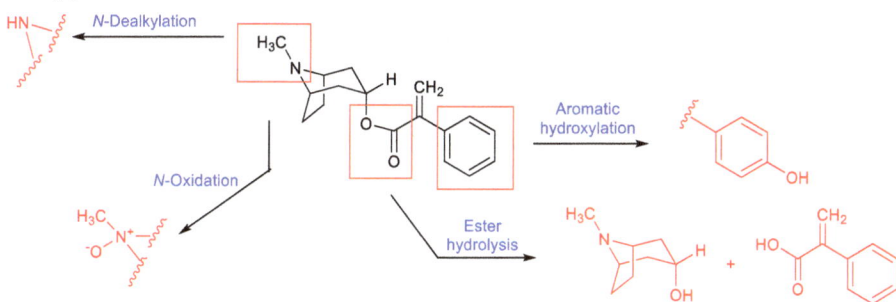

(b)

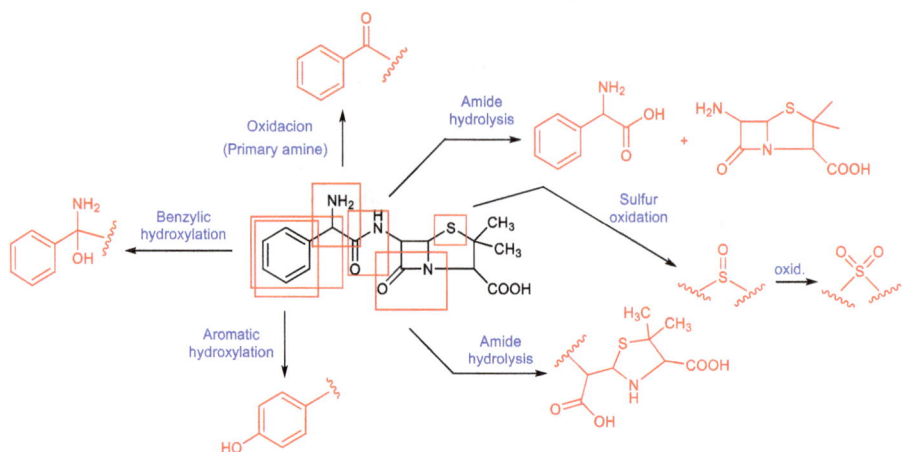

(c)

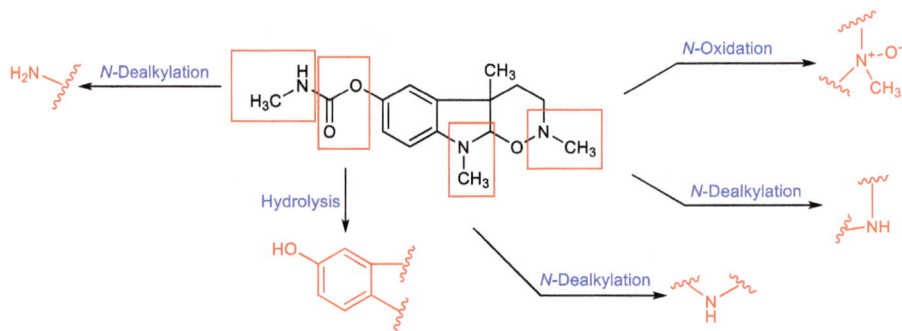

(d)

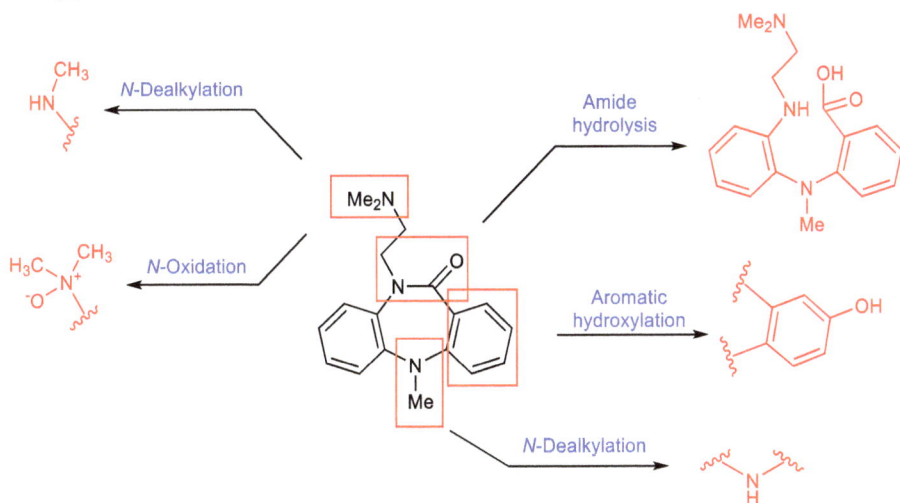

(e)

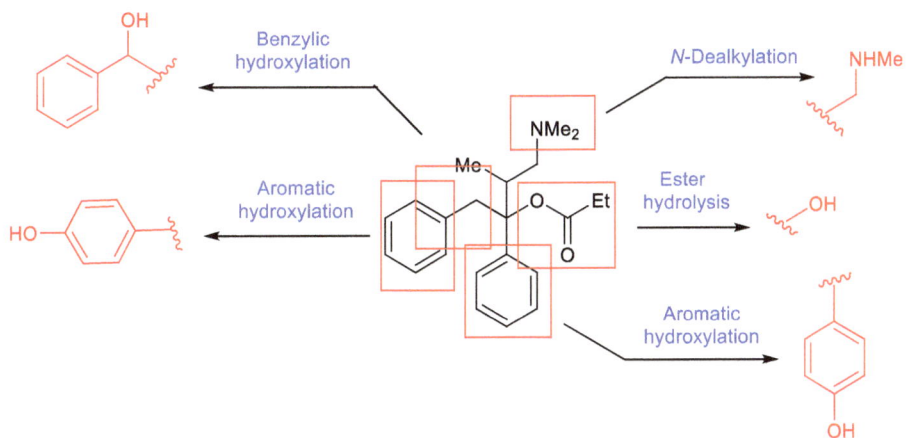

(f)

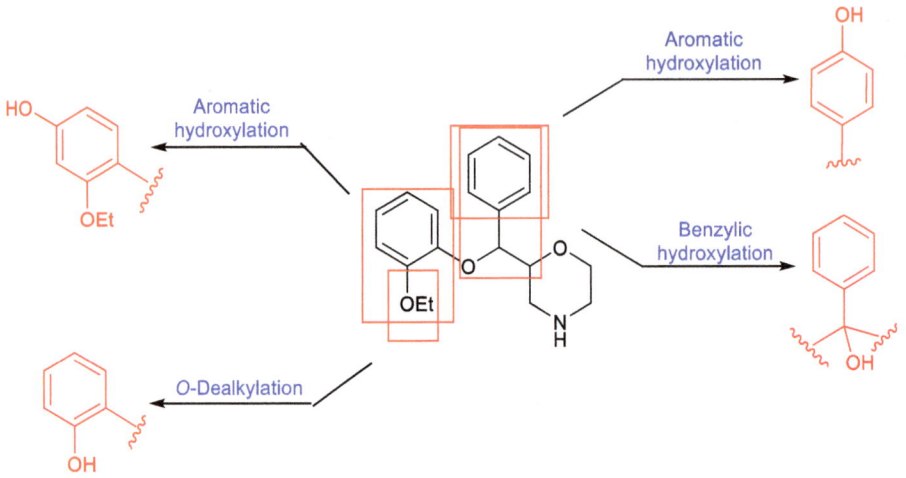

(g)

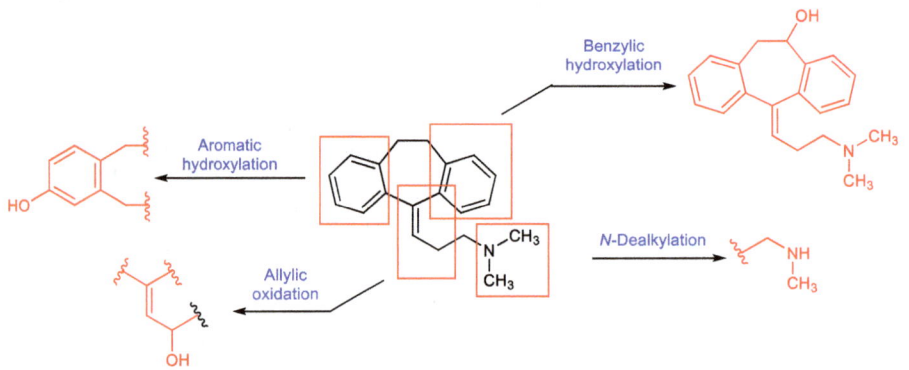

(h)

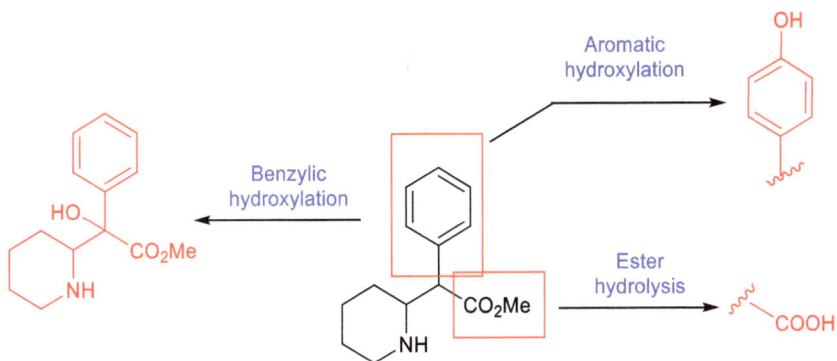

2. (a)

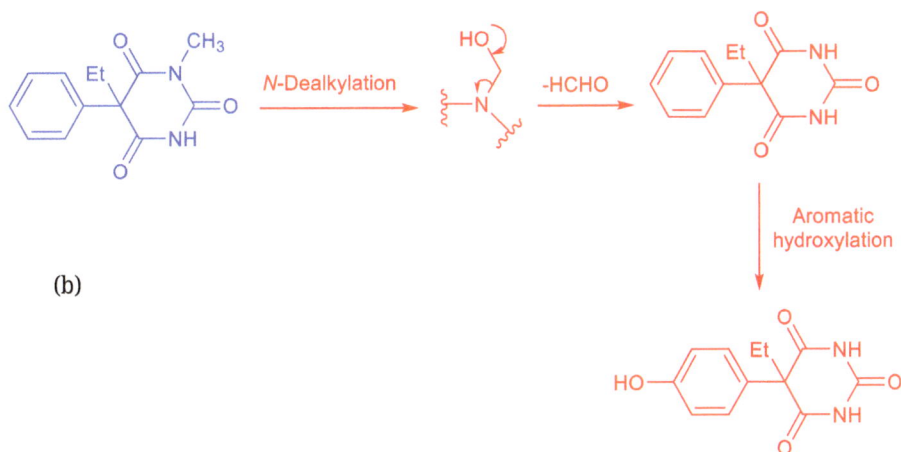

(b)

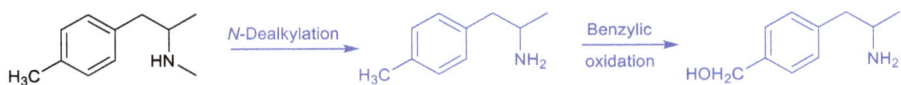

(c)

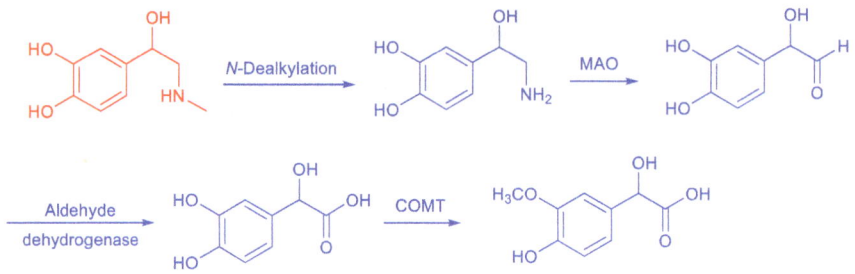

(d)

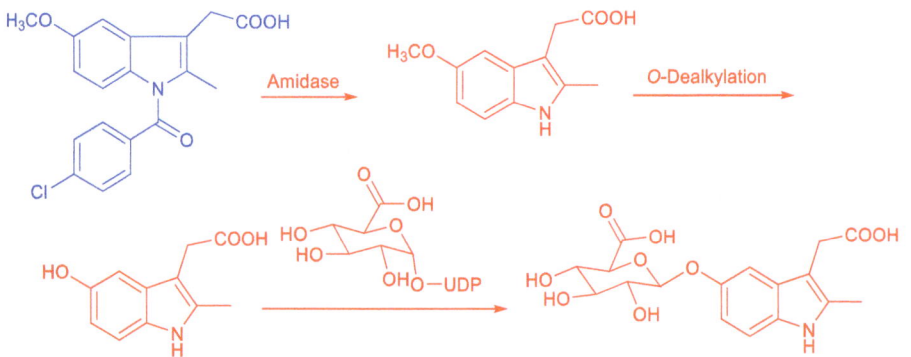

3. (a)

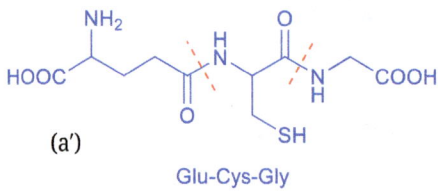

(a')

Glu-Cys-Gly

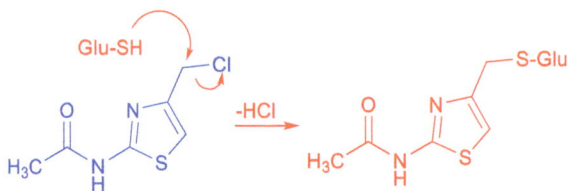

(b')

(c')

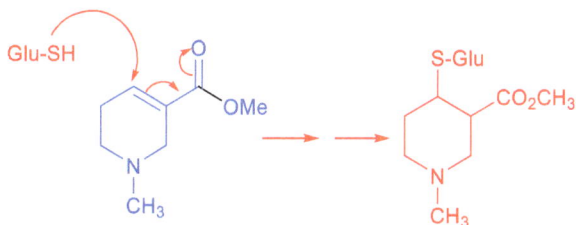

(d')

4. (a) D-N(CH₃)C(CH₃)₃ is a hard analog of D-N(CH₃)₂.
    (b) D-COOCH₃ is a soft analog of D-CH₃.
    (c) D-COOC(CH₃)₃ is a soft analog of D-COOC₂H₅.

    t-Bu⁺ cation is stable, and degradation of the ester occurs at physiological pH.
    (d) D-C₆H₄-pCF₃ is a hard analog of D-C₆H₅.
    (e) D-CH₂CONHR is a hard analog of D-CH₂COOR because the former is more difficult to hydrolyze.
    (f) D-COOCH₂N⁺R₃ is a soft analog of D-COO(CH₂)₂N⁺R₃ because the former is an unstable acyloxymethyltrimethylammonium derivative

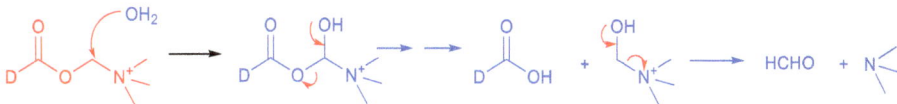

5. (a)

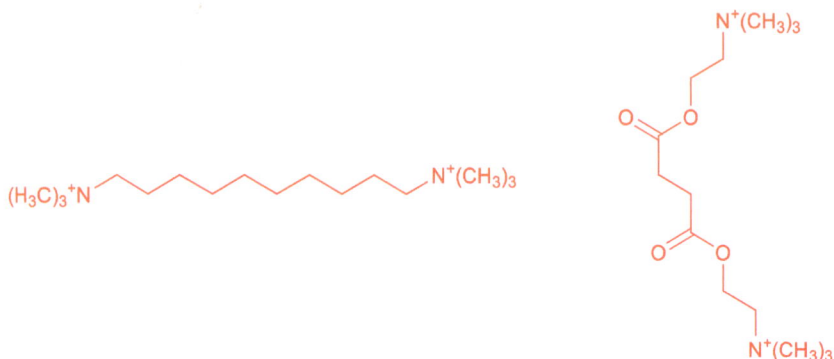

It has been transformed into a soft analog because its hydrolytic metabolism is facilitated in the modified drug.

(b)

The second drug is a soft analog of aryloxypropanolamine (with an ester group easy to hydrolyze).

(c)

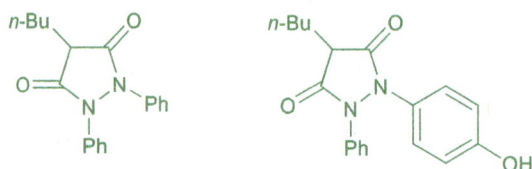

The second drug is a soft analog: it is an active metabolite of phenylbutazone.

(d)

Hard analog: the first drug is transformed into a carbamate more resistant to hydrolysis.

(e)

Hard analog: by introduction of a steric hindrance in the proximity of the ester group that hinders its hydrolysis.

6.

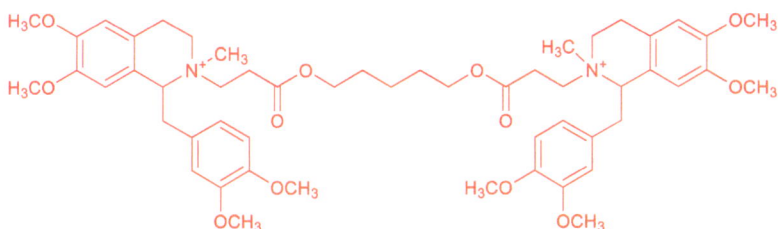

Atracurium

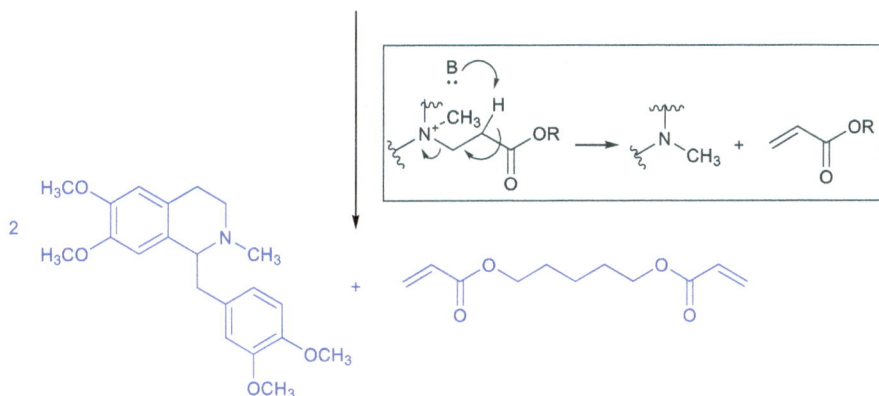

9.6 Exercises to Chapter 8

(a) (*RS*)-Fluoxetine

Fluoxetine (Prozac®, see Chapter 4 of Volume 2 of this series) was marketed for the first time in 1986. Although it has been on the market for 37 years, it remains one of the most prescribed antidepressant drugs.

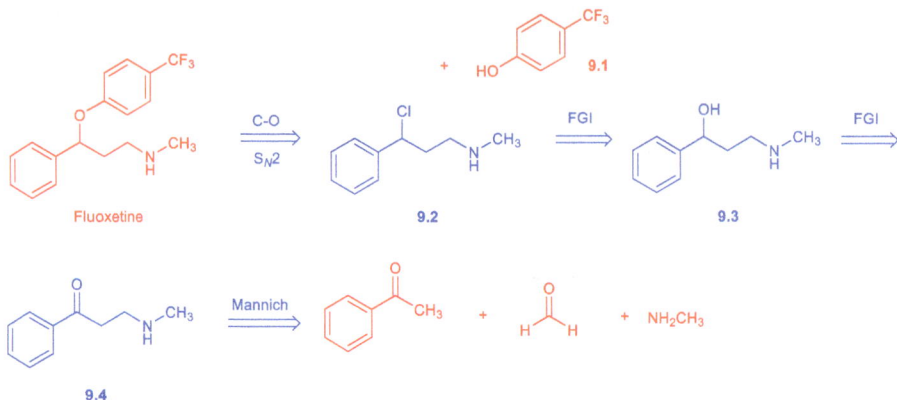

Scheme 9.1: Retrosynthetic analysis of fluoxetine.

The disconnection process begins with the cleavage of the C–O bond. The cleavage of the C–O bond leads to *p*-trifluoromethylphenol **9.1** and the functionalized amine **9.2**, the precursor of which is the amino alcohol **9.3**. The increase in the oxidation state of the hydroxyl function generates β-aminoketone **9.4**, whose disconnection process, based on a Mannich reaction, leads to acetophenone, formaldehyde, and methylamine (Scheme 9.1). Fluoxetine prepared according to the synthesis described in Scheme 9.2 is obtained in racemic form.

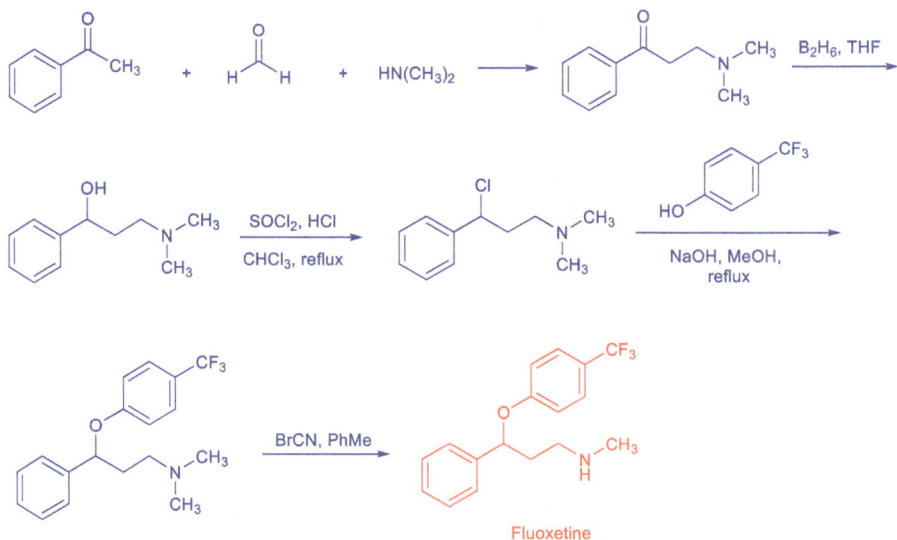

Scheme 9.2: Synthesis of racemic fluoxetine.

(b) Nifedipine

Nifedipine is a calcium channel blocker of the 1,4-dihydropyridine type, used in medicine for the relief of *angina pectoris*, as well as for arterial hypertension (see Chapter 10 of Volume 2 of this series). Scheme 9.3 outlines the retrosynthesis of nifedipine.

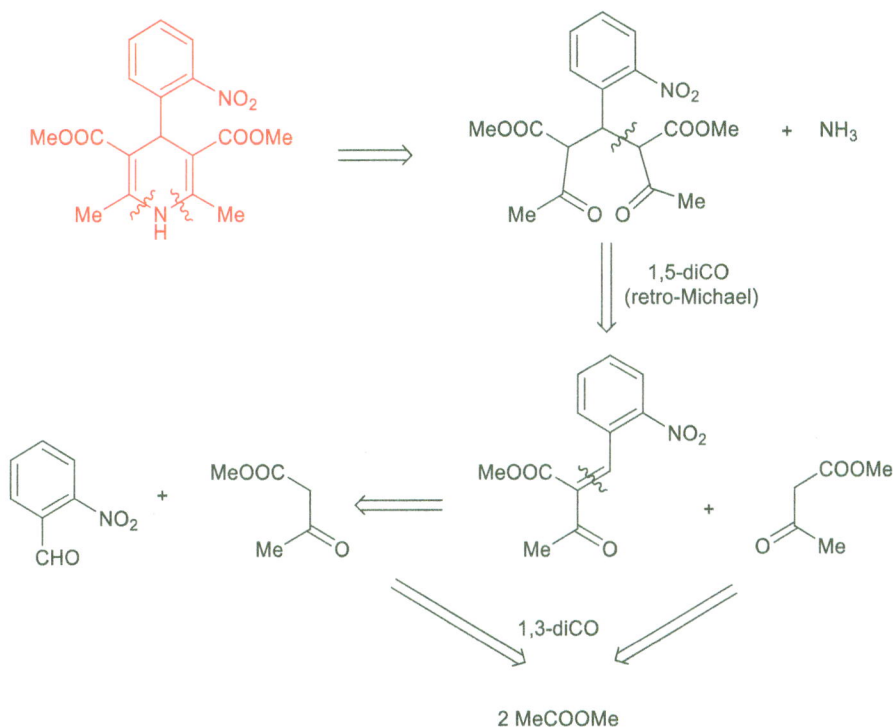

Scheme 9.3: Retrosynthesis of nifedipine.

1,4-Dihydropyridines are obtained from Hantzsch synthesis, i.e. they should be formed from 1 mol of aldehyde, 2 mol of 1,3-diCO compound, and 1 mol of ammonia. The necessary β-ketoester to react with *o*-nitrobenzaldehyde (Scheme 9.4) is prepared previously by the Claisen condensation of ethyl acetate.

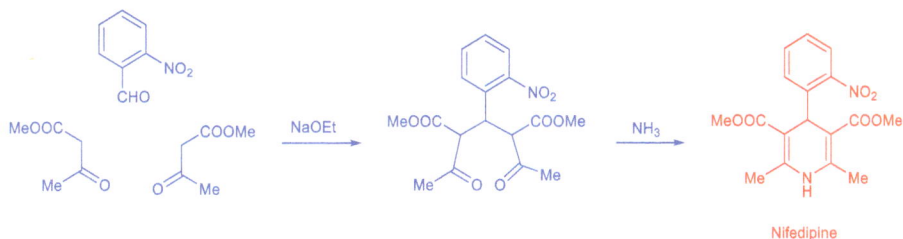

Scheme 9.4: Synthesis of nifedipine.

(i) Trimethoprim

Trimethoprim is a bacteriostatic antibiotic derived from a trimethoxybenzylpyrimidine and almost exclusively used in the treatment of urinary tract infections (see Chapter 12 of Volume 2 of this series). Its retrosynthesis is shown in Scheme 9.5:

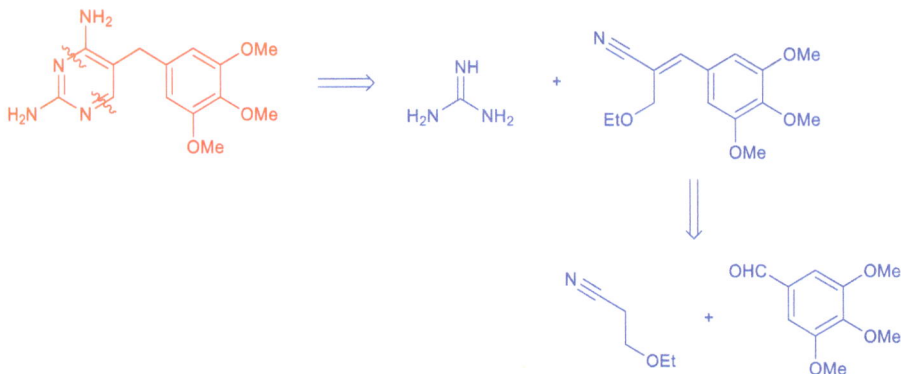

Scheme 9.5: Retrosynthesis of trimethoprim.

Synthesis begins with an aldol condensation of the benzaldehyde derivative with 3-ethoxypropionitrile. The ethoxy group in an allylic position can be displaced by a nucleophile. An amino group of guanidine is the nucleophile in this displacement process, while the second amino group is added to the cyano group. Proton 1,3-rearrangement gives rise to the drug shown in Scheme 9.6.

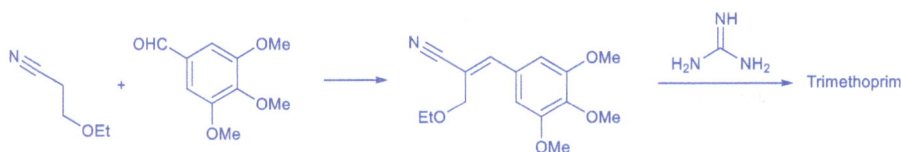

Scheme 9.6: Synthesis of trimethoprim.

(ii) Methotrexate (MTX)

MTX (see Chapter 13 of Volume 2 of this series) is an antimetabolite that has antiproliferative and immunosuppressive activity by competitively inhibiting the enzyme dihydrofolate reductase. This is a key enzyme in the metabolism of folic acid that regulates the amount of intracellular folate available for the synthesis of proteins and nucleic acids. It prevents the formation of tetrahydrofolate necessary for the synthesis of nucleic acids. Its retrosynthesis is shown in Scheme 9.7.

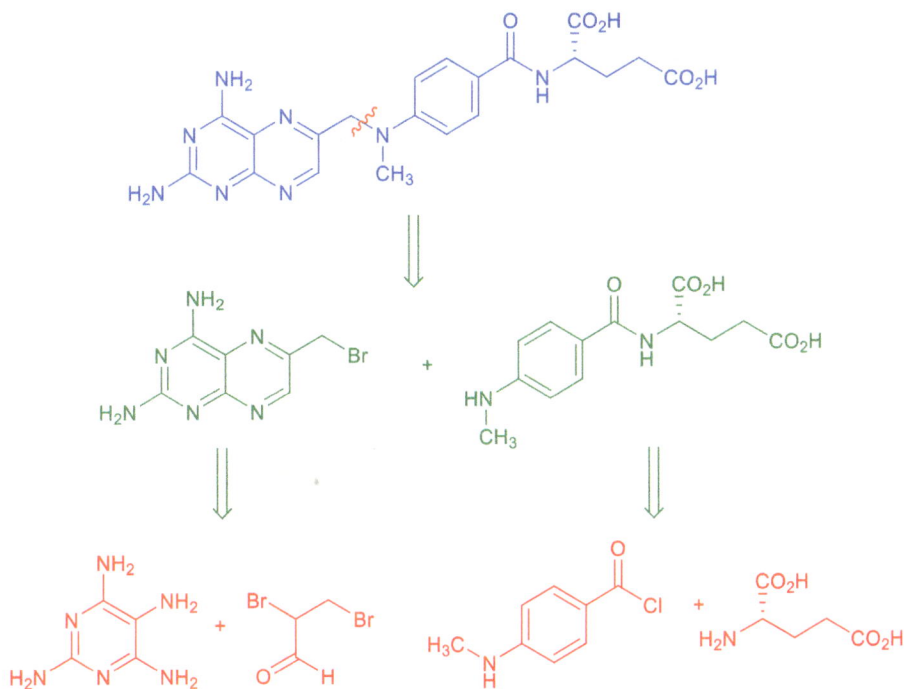

Scheme 9.7: Retrosynthesis of methotrexate.

The route begins with formation of the pyrimidine base by condensation between guanidine and malononitrile, followed by the introduction of the new amino

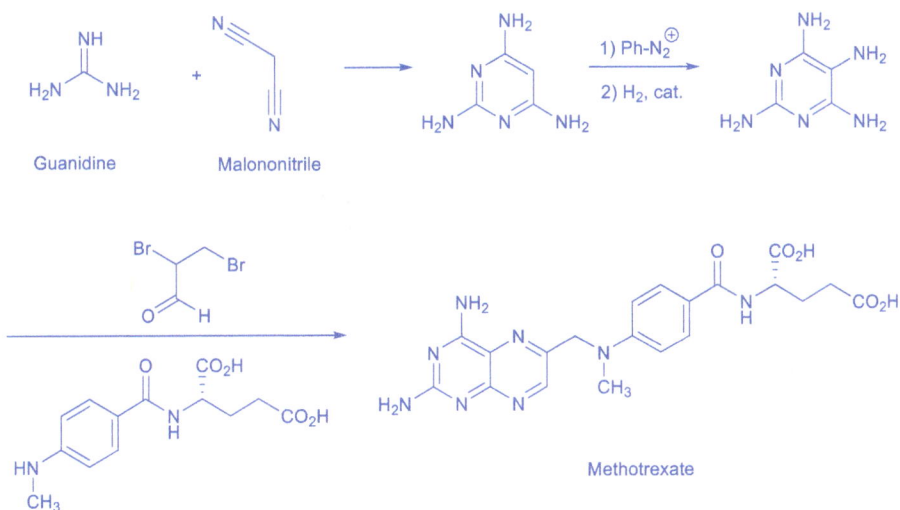

Scheme 9.8: Synthesis of methotrexate.

group by diazo coupling and subsequent reduction. The next step builds up the molecule in a single-step one-pot synthesis (Scheme 9.8).

(iii) Sumatriptan

Sumatriptan (see Chapter 4 of Volume 2 of this series) is a medicine that belongs to the group of triptans and is used for the treatment of migraine. Its retrosynthesis is shown in Scheme 9.9.

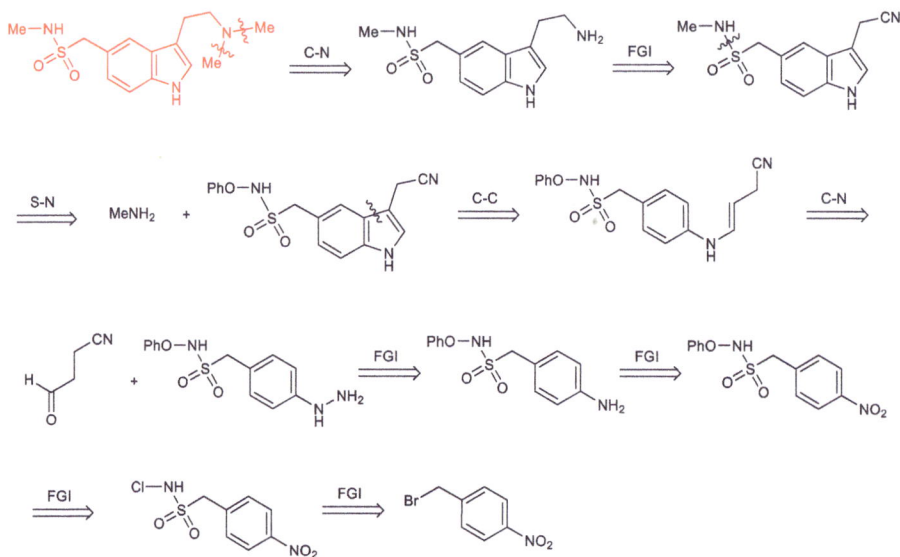

Scheme 9.9: Retrosynthesis of sumatriptan.

Synthesis of sumatriptan is shown in Scheme 9.10. Sumatriptan was the first serotonergic agonist introduced in therapeutics in 1991 by Glaxo. For its synthesis, the 4-substituted aniline by an *N*-methylsulfamoylmethyl group is the starting material. Aniline nitrogen is converted into diazonium salt with nitrous acid. Its reduction with tin(II) chloride yields the corresponding arylhydrazine, which is condensed with 3-cyanopropionaldehyde diethyl acetal to give the hydrazine, which undergoes a Fisher rearrangement to give indole **9.5**. Finally, the reduction of the nitrile to primary amine and its treatment with an excess of formaldehyde and sodium borohydride leads to *N,N*-dimethyl derivative (sumatriptan).

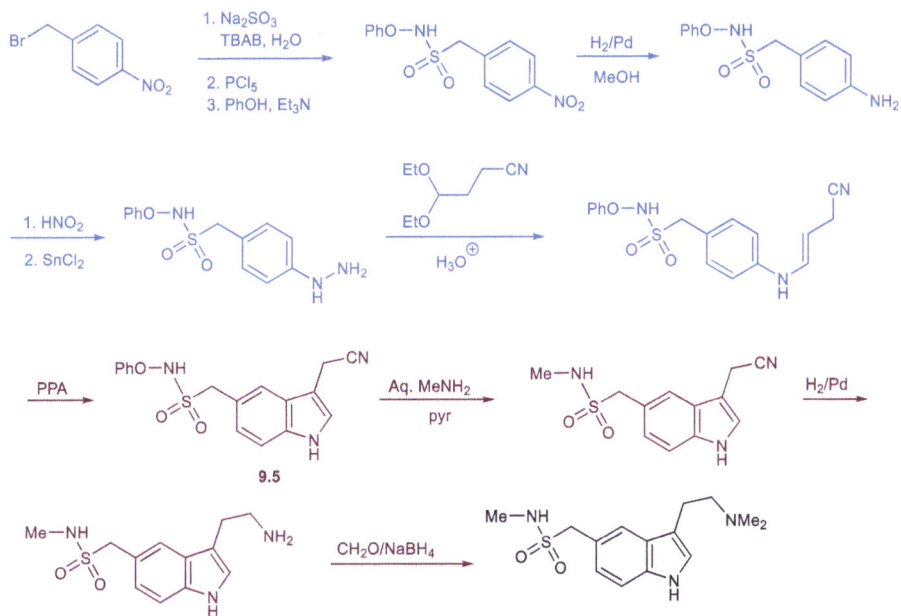

Scheme 9.10: Synthesis of sumatriptan.

Index

https://doi.org/10.1515/9783111316901-010

www.ingramcontent.com/pod-product-compliance
Lightning Source LLC
Chambersburg PA
CBHW061409210326
41598CB00035B/6152